Marry Your Customers!

Customer Experience Management in Telecommunications

Dr Janne Ohtonen

Foreword by Niall Norton

Publisher: Glamonor Publishing

Cover design: Dr Janne Ohtonen, Miika J. Norvanto, Rob Williams

Email: janne@threecustomersecrets.com

LinkedIn: http://www.linkedin.com/in/janneohtonen

Book website and free resources:
http://www.threecustomersecrets.com/member

ISBN 978-952-68055-3-5 (PAPERBACK)
ISBN 978-952-68055-4-2 (PDF)
ISBN 978-952-68055-5-9 (EPUB)

ABOUT THIS BOOK

---

*"I, Telco Leader, take thee, Precious Customer, to
be my Contracted Partner, to have and to hold,
from this day forward, for better, for worse, for
richer, for poorer, in system failure and in excellent
network coverage, to care and to cherish, till churn
do us part, according to T&Cs binding ordinance;
and thereto I pledge myself to you."*

Telecommunication service providers have traditionally competed with technology. While technological advancement will continue to stay highly relevant for the industry, the competitive front line has shifted towards customer value and experiences. This **'Marry Your Customers!'** book gives customer relationship counselling on:

- **revealing the naked truth - that customer experiences directly impact the bottom line for telcos;**
- **using a practical customer experience management framework tried in better and in worse; and**
- **how to use technology to increase customer value and to ensure long-term relationships.**

The Customer Experience Counsellor, Dr Janne Ohtonen, has delivered various challenging customer experience transformation programmes over the past two decades in the telecommunications, travel and retail sectors, several of which included double-digit performance enhancement for the businesses. He now, together with other telco leaders, shares with you the insights and undisclosed strategic approaches that are successfully used by various telecommunication and other organisations around the world today.

**THE BOOK INCLUDES FREE ONLINE MATERIALS AT
http://www.threecustomersecrets.com/member**

DEDICATION

To those persistent partners, who will not shy away even from difficult circumstances, to add more value to both customers and to the business.

"Everything starts with the customer"
- Louis XIV, King 1643-1715, France

LET'S GIVE A BIG HAND

"Great things happen to those who don't stop believing, trying, learning, and being grateful."
- Roy T. Bennett, Author, The Light in the Heart

I want to thank my family, who have supported me for decades on this road of developing customer experience management methods. The information shared in this book has taken thousands of hours to trial and countless nights away from home. Big thanks to my family and friends for your support and understanding. I am much obliged to **Hanna Norvanto-Ohtonen, Daniel Ohtonen, Arja Ohtonen, Jarmo Ohtonen, Mira Ohtonen, Mikael Ohtonen, Minna Karjalainen, Mikko Karjalainen, Toni Karjalainen, Nea Karjalainen, Jukka Norvanto, Miika J. Norvanto, Liisa Norvanto, Elisa Norvanto, Pirita Norvanto, Mikaela Norvanto, Miral Ismail and Janne Saarinen**.

Special thank you to **The Association of Finnish Non-fiction Writers, TalkTalk Business UK, Tele2 Russia, Black Lion Pictures** and **Openet** for sponsoring this book.

I would like to extend my gratitude to **Niall Norton, Tommi Hännikkälä, Tuukka Heinonen, Barry Marron, Enri-k Salazar, Nina Gyubbenet, Danny Sullivan, Jeff Bezos, Warren Buffett, Tony Robbins, Robin Sharma, Stephen O'Loughlin, David McGlew, Sean Broderick, Martin Morgan, Pete Tomlinson, Dan Richardson, and Andrew Sweeting**.

A big thank you to all those people who write useful books for others to learn from. Finally, the greatest acknowledgement goes to my Lord and Saviour, **Jesus Christ**, who makes all this possible.

WEDDING MENU

II. CUSTOMER EXPERIENCE MANAGEMENT FRAMEWORK FOR BETTER AND FOR WORSE..65

III. A USER MANUAL FOR ALIGNING THE TECHNOLOGY TO SUCCESSFUL OUTCOMES...121

BEST MAN'S SPEECH, BY NIALL NORTON, CEO, OPENET

The topic of Customer Experience Management (CEM) is often misunderstood by those who have not had an opportunity to understand what it really means, and why it is valuable. This book, while focused on the telecommunications industry, contains many truths that apply to every business. And this makes this book important as it provides great insights, and it facilitates the reader's mindset in an approachable manner that explores CEM in a very pragmatic fashion.

Telecommunications is a tough market. Over the past 25 years there have been innovation cycles for services, devices and technology in the vast growth area of allowing people and businesses to communicate more efficiently with each other, it helped share new ways in which people engage with each other, and it allowed the internet to become part of everyday life. In turn, this has increased happiness, relevance and productivity.

For different services, and for various audiences, telecommunications service providers have enjoyed a lifecycle that starts with almost monopoly market dynamics and ended with commoditization. A small number of highly capitalised, licensed, businesses grew the importance and relevance of the sector to unimagined success by providing services that their customers wanted in the most basic area of being able to communicate more efficiently than before.

For many years, the most important features of the services that a telecommunication service provider delivered to its customers were the reliability of the communications and the usability of the devices. These differentiators were for a very long time built on avoiding less than almost perfect systems and high reliability that often-stifled rapid innovation. The consumer, or the business personality of the end-customer (user), was sometimes ranked as third in the scheme of important things.

Over time price competition and the improvements in technology allowed the telecommunications service providers to widen the portfolio of services they provided to their customers. This permitted pretty much ubiquitous and high-speed internet connectivity to almost all customers, and this brought the telecommunications service providers into competition with businesses like their own, and now also with internet service providers.

Internet service providers had evolved in a different arena – one where service delivery was not within their control and therefore was not a differentiator. Price, relevance, and usefulness were the prime elements of the value propositions that emerged for the most successful Internet service providers. The winners had to both know their customer and control the customer's experience as the competition was fierce, and barriers to entry to this market were (and are) low.

Rapid innovation and deep understanding of the end-customers experiences were the key differentiators. Pivoting and segmentation of profiles remain a key attribute for the internet players – the value being emphasised in areas of service consumption and not on the technology of how the service is delivered. Therefore iPhone, iPad and Android killed Nokia devices.

When the internet and telecommunications markets intersected, the internet service providers were very well placed to engage customers and own the relationships, and the telecommunications service providers were slow to adapt to this change. As a result, they have been under significant pressure to retain the value of their place in the value chain in communications – as the end devices and customer engagement channels were taken by other players.

Left with a limited set of tools to compete with for revenues, and perceived parity of connectivity coverage as networks were built out, the telecommunications service providers were left really only with price to compete with in the market. So, revenues declined and the pressure to reduce costs to follow the internet service provider models increased dramatically. But you cannot save your way to greatness – and endless price competition terminates in the graveyard for businesses.

This is all depressing stuff if you are invested in the telecommunications market – whether as a service provider, a vendor or a professional. But happily, it is not the whole story or the end of the story.

Many of the better telecommunications service providers understood in time the dynamics of what was underway and have sought to change their models to being more enabled to compete or co-exist with the internet service providers. There are a number of effective strategies that have been developed by them, including: collaboration with others and bundling of services, creation of their own content, reduction in the complexity and the costs of network management, and perhaps most significantly the development of an appreciation for the value of truly excellent Customer Experience Management.

CEM, as outlined in this book, has huge and tangible business value in terms of churn reduction, loyalty and referrals, increased spend in the telecommunications channel and services tailored to the markets of each customer – all of which increase profitability. The enabling technologies allow telecommunications service providers to deliver on CEM in ways just not possible a few years ago. And things are changing. I am an optimist and hope that this book will further help with this change by helping shape a growth mindset in the CEM area.

CEM is not a strategy for a business – it is a cultural aspiration. It changes the end-customer perception, creates trust and helps make sure that the provider of services is providing value. It is built on customer understanding and knowledge. And it means that the providers of services must understand what a customer actually values and finds relevant. This type of relationship cannot be automated or faked – it must be sincere and consistent. This book sets out, far more elegantly than I can, how to achieve this goal for a service provider, and I believe that this will be fundamental in the telecommunications industry looking forward.

I started my journey in business in the finance area and flirted with technology and other "hardened" disciplines before I truly appreciated the importance of the customer experience and how this translates to real economic value. But, once understood, it is impossible not to see its importance is all industries and endeavours. 'Marry Your Customers' is a pretty accurate title for this book – it certainly helped me in marrying the mindset.

Niall Norton

Niall Norton
CEO
Openet

WHAT ARE WE HERE FOR?

"We see our customers as invited guests to a party,
and we are the hosts. It's our job every day to make
every important aspect of the customer experience
a little bit better."
- Jeff Bezos, Founder, Amazon

WHAT AN EARTH IS CUSTOMER EXPERIENCE MANAGEMENT SUPPOSED TO BE?

"It takes 20 years to build a reputation and five minutes to ruin it."
- Warren Buffet, Money Guru, USA

Reminding you of the many stories that are known well by all of us around the world on how customer experiences can affect business is probably not necessary. Most of us remember how United Airlines first broke guitars[1] and then committed "domestic violence" by throwing customers out from their planes[2]. I am sure you have many stories of your own (especially if you happened to be a Vodafone customer in the recent years[3]) of things gone wrong with your customer experiences.

But that's not the point. The point is that any marriage can be broken with enough abuse from either of the partners. In divorce, you may end up losing half of what you have, but in a break-up with the customer you lose it all! And no worries, this isn't just marketing talk to scare people off like North Korea's alleged nuclear program[4], but comes with a tangible impact on your revenue (more on that in the next chapter on the financial impact of the Customer Experience Management).

There are various definitions of Customer Experience Management (CEM) as we will soon discuss, but if we boil down the purpose of CEM, it comes to:

---

[1] https://www.youtube.com/watch?v=5YGc4zOqozo

[2] https://www.youtube.com/watch?v=VrDWY6C1178

[3] Vodafone fined in the UK https://www.ofcom.org.uk/about-ofcom/latest/media/media-releases/2016/vodafone-fined-4.6-million and in Ireland http://bit.ly/vodafone-250k-fine

[4] http://www.express.co.uk/news/world/855886/World-War-3-US-North-Korea-Trump-Kim-Jong-un-missile-launch-Japan

- **RETAINING existing customers in a sustainable way**
- **ACQUIRING new customers in a sustainable way**
- **Maximising operational EFFECTIVENESS through LeanCEM**

I understand it may come as a shock not to see things like 'increasing the number of satisfied customers' and '*improving the Net Promoter Score*' on the list. But this is the thing: they are a means to an end. And the purpose of the CEM should be SUSTAINABLE GROWTH for the business delivered through optimised OPERATIONAL EFFECTIVENESS. This book is full of hard messages, but this actually may be one of the hardest ones. Especially for the top leadership. CEM is not about creating short-term profit by any means necessary. It is not only about this or the next quarter. It is about ensuring the business is healthy today, tomorrow and the next year (not to mention the many years after that); EVEN IF IT MEANS LESS MONEY TODAY!

And that's hard to swallow. I get it. If the firm measures you by quarterly and financial year results, then the long-term focus is difficult. But it just has to be done. Otherwise, there's no future for the business in the long-term. And that is where CEM can help you. It gives practical tools for ensuring the company will stay relevant for its customers for a long time to come.

Is forgiving others challenging for you? If it is, you are not alone on that[5]. And that is also what Warren Buffet is referring to in the above quote. It's hard to get the customers to come in, and when things go wrong, they get angry quickly and forgive slowly. It's painful really. So, what we use CEM for is to optimise the likelihood of a positive experience and deal swiftly and appropriately with situations where the experience goes wrong. CEM is all about the customer love, so take your rainbow banners out, you'll need them!

From the perspective of the whole of society, such short-sightedness has a massive impact on our daily lives and happiness. Companies that are unwilling to deliver the value they produce in the best possible way are creating Moments of Miseries for people who are victims of such practices. Just have a look at the faces of people at the Victoria Station in London when they exit the Sothern

---

[5] http://www.bbc.co.uk/news/magazine-13154300

Railways trains. You won't see any smiles being exchanged there. And no wonder, Southern being the most unpunctual train operator in the whole country with one of the lowest customer satisfaction scores in the UK. But they meet their business KPIs, so what's there to moan about for their managers? Bloody customers … getting in the way of doing business!

On a more serious note, the academic circles define Customer Experience Management as the means of creating advocates for an organisation's brand, products, and services. This includes turning customers from the merely satisfied to vigorously loyal and then, in today's social media-aware age, from hardly loyal to active advocates. CEM is all about building an EMOTIONALLY weighted relationship with EACH customer. This is well aligned to the previously presented three-point purpose of the CEM.

An industry publication, Heavy Reading[6], defines *"a CEM system as one that collects data related to Customer Experience (CEX) from multiple sources and then models and analyses CEX data and recommends actions by CEX [pronounced as /sɛks/] analytics"*[7].

Sure, data is needed to do this, but we need a more holistic view of CEM, one that aligns customer needs to business strategy. That is an understanding of why we want to produce added value, what we have to do and how it will be done. This includes, but isn't limited to, data and insights, thus rendering Heavy Reading's view too narrow to our taste.

Ironically, according to Ovum, there's no consensus in the telecommunication vendor community on what Customer Experience Management is[8]. I suppose that is not surprising as the competition for the CEM dollars available from the service providers is fierce. Everyone wants to create their unique selling point (USP) and market differentiation (MD for those with TLA[9] fetish).

This leads to a wide selection of wildest definitions of the CEM, each of course conveniently matching whatever the specific vendor is offering. Such an approach by vendors does have a little entertainment value (if you are into such humour), but it doesn't help

---

[6] http://www.heavyreading.com/document.asp?doc_id=38129

[7] Addition in the brackets by the author. Quote source:
https://www.vanillaplus.com/2015/09/08/11387-cem-can-csps-avoid-a-race-to-the-bottom-on-price/

[8] Ovum - CEM or CRM? The Puzzle for Telcos, Product code: IT024-000015

[9] TLA = Three-Letter Acronym

the service providers to focus on what CEM indeed is and what they should do about it. From a service providers' perspective, the best thing that vendors could do for them would to be very clear on which of these CEM areas they can help the service providers with:

- **Brand, communications, marketing, and sales**
- **Customer service and proactive problem prevention**
- **Value adding products and services**
- **End-to-end customer journeys leading to successful customer outcomes**

CEM in the company context will be discussed extensively throughout this book. We all know that customer retention is a high priority for telcos. Traditionally, communication service providers have been perceived as offering poor customer service, but this is now starting to change as they begin to acknowledge customer experience optimisation as one of the biggest differentiators. The race to improve client retention, operational efficiency and to drive new revenue streams through CEM has begun!

Delivering an ideal customer experience can drastically bring down churn rates[10], increase ARPU/ARPA[11], and reduce operational costs as we will soon see. Digital service providers need to look at the managing of the customer experience as more of a methodology or a practice than a product category. We will discuss this in more details later in this book.

And what about you? There is a whole new CEM industry that advises top management on how to utilise customer experiences as a growth strategy for the business. You could be one of them. I have seen zero to 6-figure salary growth in less than five years with one or two promotions per year for people who go with this knowledge to their leaders. And for those who work in marketing, customer service or in IT, customer-centric approaches have become mandatory.

If you want to prove your place in the esteemed bunch of CEM specialists, you can even get yourself a CCXP[12]-certification. I can't say what the benefits will be for you specifically, but I can promise that following this path won't leave you empty-handed. Going back to what CEM is, from your perspective it helps you to become a

---

[10] CEM Strategy: A Must for Telcos, Pipeline Magazine

[11] Average Revenue Per User (ARPU) and Average Revenue Per Account (ARPA)

12 Certified Customer Experience Professional, http://cxxp.org

happier employee (or leader) as you won't have to deal with so many customer complaints and issues. And it will also make your life happier as you receive better experiences from the brands you regularly interact with (assuming they care about your experience).

We have already discussed what Customer Experience Management is from various perspectives. And maybe it has felt a bit fluffy at times? If you like hard, cold numbers, then you'll love this! There is a way to give a score for customer experience performance. The equation goes as follows:

$$CEX^{13}\ Performance = \frac{Experience\ Performance}{Expected\ Experience}\%$$

This CEM equation shows the ratio of perceived customer experience performance against the expected experience. The actual numbers for the CEM equation are highly dependent on what kind of data you have available. Let's assume that we can use the scale from 0 to 10 for both of the above data points and we have the data available from our Voice of the Customer programme. In that case, our CEM equation for an individual customer (or if using averages, from our customer base) could look something like this:

$$CEM\ Performance = \frac{6}{9} = 67\%$$

The optimal CEM Performance is 100% as that's when the perceived experience and expectations are balanced. Anything below 100% is under-delivering, and anything over 100% is over-delivering (and potentially losing money on doing that). In the above example we could say that our business is under-delivering the customer experience for that specific customer (or if using averages, for the customer base) by (100%-67%=)33%.

It is known that CEM is the top priority for operators, with 68% of telco leaders citing it as the number one strategic priority for their organisations, while 82% viewed it as a top-three consideration over the next three years, according to Ernst Young[14]. And what do telecom leaders expect out of these Customer Experience Management initiatives? Genesys[15] conducted a study on this and

---

13 CEX = Customer Experience

[14] Ernst Young: Global Telecommunications Study: Navigating the Road to 2020

[15] https://www.superoffice.com/blog/customer-experience-statistics/

gave us these figures:

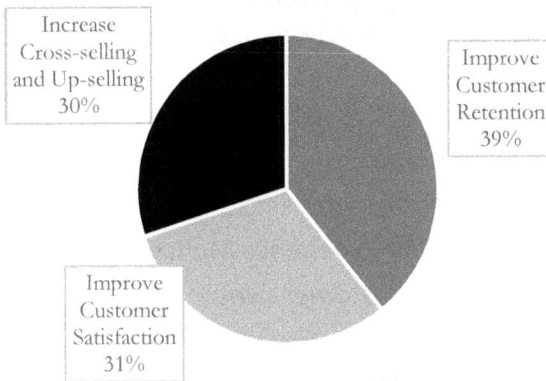

You probably spotted already few differences as well as similarities in the graph above to what we have discussed earlier. The main reason for telcos to improve customer retention would be to drive growth for the business. Same applies to customer satisfaction and sales related efforts. But the above study has missed the point of utilising the LeanCEM[16] approach to drive cost-effectiveness. In short, LeanCEM focuses on optimising a company's operations based on delivering value to the customers (more on that later in this book). It includes taking out non-value-adding activities. Maybe we will see this perspective added in the future studies?

Another way of looking at what CEM is in the telecoms sector is through what the leaders are investing in currently. Here's data from the IQPC 2017 telecom CEX leaders survey[17]:

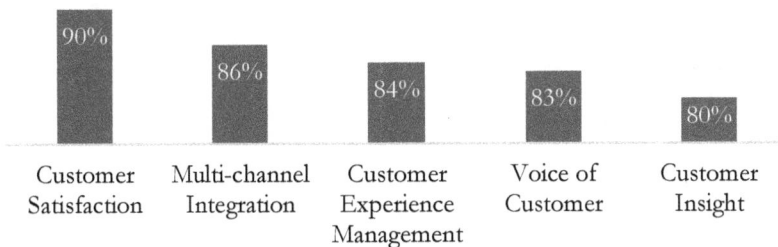

---

[16] Instructions for practical application of LeanCEM are available in 'The 5-Star Customer Experience' book by Dr Janne Ohtonen

[17] IQPC: Telecoms Report 2017 – CX Investment

We have now discussed the CEM from various perspectives, including social, company, personal, and customer. It is time to conclude what the Customer Experience Management in Telecommunications is:

"Customer Experience Management
is a customer-centric and
Sustainable Growth Strategy
for the business."
- Dr Janne Ohtonen, Author

WHAT CEM IS NOT!

If this book was about something other than Customer Experience Management, then the previous chapter on what CEM is could suffice. But when it comes to customer experience, there are many opinions on what it is (as discussed) and what it is not or should not be. Here's a non-exhaustive list of things CEM is not:

- **Business Strategy**
- **Voice of Customer (VOC)**
- **Customer Journey Map**
- **Customer Relationship Management (CRM)**
- **Marketing and Sales**
- **User Experience (UX)**
- **Customer Service**

I have a feeling that you are already all too familiar with discussions that go something like this: *"CEM is same as marketing"* or *"Our sales are already customer-centric and do CEM"*. Well go and ask those salespeople to define the CEM they do and you will start to see that they may touch part of CEM, but are unlikely grasping the breadth and depth of Customer Experience Management. Anyway, for your future discussions, here are main key points on why CEM is not any one of the above things.

BUSINESS STRATEGY is a proposal for action designed to achieve a set of goals that a company has set for itself. It says how business should be done to meet those desired goals. Thus, a business plan may or may not be customer-centric. Business Strategy could state that *"I, as a business, acknowledge that the customer is my lawfully wedded partner and I will remain loyal to my customer."* Companies that foster this thought in their Business Strategy in one way or another will be much more successful than those who do not. However, Business Strategy is not sufficient to cover CEM alone.

VOICE OF CUSTOMER is a way to listen and understand customers. It is an essential contributor to any reputable CEM programme. But again, it alone is not same as CEM. It can give insights into how to achieve business growth through a CEM approach, but it will not fulfil the whole definition of CEM that we did in the earlier chapter by itself.

CUSTOMER JOURNEY MAPS are in the same position with VOC programmes. They are a crucial part of a successful customer experience management, but they are not sufficient alone.

CUSTOMER RELATIONSHIP MANAGEMENT is often mistakenly seen as same as CEM. However, CRM is on the same line with VOC and Customer Journey Maps. It is an essential contributor to customer communications, but it is not holistic enough to be a CEM approach[18]. To elaborate a bit more on this point, CRM and CEM are like the bride and groom. You can't have one without the other[19]! They complement each other:

CEM	CRM
Emotional Value	Functional Value
Focuses on the company's value to customer	Focuses on the customers' value to the company
People and interactions in the centre	Systems and transactions in the centre
"*Right brain*" feelings	"*Left brain*" engineering
Customers pull value	Company pushes communications

MARKETING AND SALES should report to CEM, not the other way around. As discussed earlier, CEM is a growth strategy for the business, and it has a tangible impact on the bottom line. Thus, marketing and sales are tools to implement certain aspects of CEM, but they are not the answer to all the questions that CEM poses.

USER EXPERIENCE is about the best possible experience through whatever channel customers use. Again, it is an essential contributor to CEM success, but it is too narrow a view to cover the whole CEM domain. In proper CEM strategy user experience is connected to customer journeys and the Customer Experience Blueprint. Some people call such an approach as Design Thinking. For us, even Design Thinking is a tool in the CEM domain.

Finally, CUSTOMER SERVICE, everyone's favourite. I have heard it dozens of times and will hear it again, how CEM is same as customer service. Nope, it is not. Various reports highlight the difference between the two (previously mentioned Ovum report as

---

[18] Ovum report: CEM or CRM? The Puzzle for Telcos

[19] Al Bundy – Love and Marriage

one example). Customer service is again a component within the CEM domain. It helps an organisation to respond to customer queries and proactive contact customers on various issues. CEM directs how these efforts in the customer service function are carried out.

WHAT WILL YOUR BUSINESS GET OUT OF CEM?

"The customer experience is the next competitive battleground."
- Jerry Gregoire, ex-CIO, Dell

If you are married, why did you get married? Or if you ever will get married, why would you do it? Would you assume an improvement in a few areas of your life or all area of your life? I would hope my life would get better in all areas, but as that may not be possible, then in overall. So it is with CEM. There are many strategies, methods, and tools out there that can deliver in one area or another. Only a few approaches can give encompassing business benefits. This chapter is just a sneak peek into the vast benefits that are available from Customer Experience Management. We will discuss these benefits in much more details in the next chapter **I. NAKED TRUTH: FINANCIAL IMPACT OF CUSTOMER EXPERIENCES**.

The inventor of Net Promoter Score, Frederick F. Reichheld, once said: *"On average, the CEOs of U.S. corporations lose half of their customers every five years."*[20] The truthfulness of that statement is debatable. However, one of the biggest benefits of the CEM can be to prevent churn from happening. We defined CEM to be a growth strategy for the business, but in some cases it is much more than that; a lifeline, as otherwise the firm would go belly-up. The benefits of Customer Experience Management come in several forms:

- **Strategic,**
- **Financial, and**
- **Operational**

---

[20] Source: http://bit.ly/ffr-quote1

The current misalignments in telecommunication sector are negatively impacting revenue, cost and customer service opportunities at a higher level than ever before. Much of this is caused by outdated thinking and legacy technology. The market environment where digital service providers operate has evidently changed, and there are several levels of corporate misalignment that impair the business. Even those firms that have been optimised several times over with approaches like Lean Six Sigma are suffering from these challenges, which show themselves in many forms:

- **The velocity of business optimisation methods being too slow. There is no time for 2-5 years digital transformation anymore.**
- **Continually changing customer behaviour making companies quickly irrelevant. Also, change of regulations has become much faster, forcing digital service providers to change how they operate.**
- **Imperfect product and service capabilities are leading to high customer churn. The cost of this for telecommunication sector globally is over hundred billion dollars per year.**
- **Inert company structures are hindering quick change. Times when doing the same old thing was safe are gone.**
- **The rapid pace of change exhausting business resources. Most companies don't have hundreds of millions of dollars to invest in replacing their legacy systems.**

Many operators are sitting on a redundant cost base and not even aware of missed revenue opportunities. A new kind of customer-centric, design thinking is called for, building on Customer Experience Management strategies. CEM helps the businesses to become more customer-centric, easier to deal with and to produce a higher return for all stakeholders.

The strategic importance of CEM is utmost. It contributes to competitive advantage, market share, the share of wallet, and brand consistency of a firm. But don't take it just from me. Based on a study carried out by Ernst Young[21], this is how important it is:

---

[21] Ernst Young: Global Telecommunications Study: Navigating the Road to 2020

68% **Of telecom operators view customer experience as their number one strategic priority!**

The financial importance of CEM will be discussed in more details in the next chapter and throughout the book. The most prominent contribution comes from increasing customer loyalty and retention, which will both impact the bottom line for the business. Just to highlight the potential money you are leaving on the table, here's data from Forrester Research[22]:

INDUSTRY	ANNUAL INCREMENTAL REVENUE PER CUSTOMER	AVERAGE NUMBER OF CUSTOMERS PER FIRM	TOTAL REVENUE TO GAIN WITH CEM
Wireless Service Providers	$3.39 x	82 million =	$278 million
TV Service Providers	$6.11 x	17 million =	$104 million
Internet Service Providers	$5.26 x	16 million =	$84 million
Telecom Total	--	--	**$466 million**

Though you need to take the above data with a pinch of salt (as they have made several of assumptions we don't know to get these numbers), there is still significant revenue to be gained. And this is excluding the impact of churn reduction and advocacy.

Operational importance of CEM is something that is not well understood or documented. In the next chapter we will go through a mathematical model that shows the potential for improving operational effectiveness through CEM. Northstream[23] estimated that

---

[22] Better Customer Experience = Higher Revenue Growth Forrester Links CX To Greater Loyalty And Revenue Growth Across Industries

[23] Quantifying the value of Omni-channel CRM for Telecoms by Katie Matthews

just implementing a part of CEM strategy (i.e. Omni-channel optimisation), in cash terms represents an opportunity of \$325m all the way up to \$1.6m in annual industry savings just in Western Europe.

To realise this potential, a LeanCEM approach needs to be adopted where the operations are optimised from customer-centric perspective. Putting the customer in the centre of operations will have a huge positive impact on everyone's focus and their priorities in business. By having customer data guiding operations, it is possible to ensure that the customers stay happy and your business stays relevant.

As a summary, the more you put into Customer Experience Management, the more your business will get out of it. Assuming you work to right matters.

MAY THE FORCE BE WITH YOU, CEM

"Customer Experience Management,
the force will be with you. Always."
- Obi-Wan Kenobi, Star Wars
(liberally paraphrased)

Previously we discussed the benefits of marrying with customers. In this chapter, we will look more into the forces that impact CEM. Yes, we are talking about the in-laws and the stupid cousin you have to tolerate every Christmas. CEM is an overarching concept that touches everything a business does with its clients and more. Like marriages, customer experiences have a wide variety of different forces impacting them:

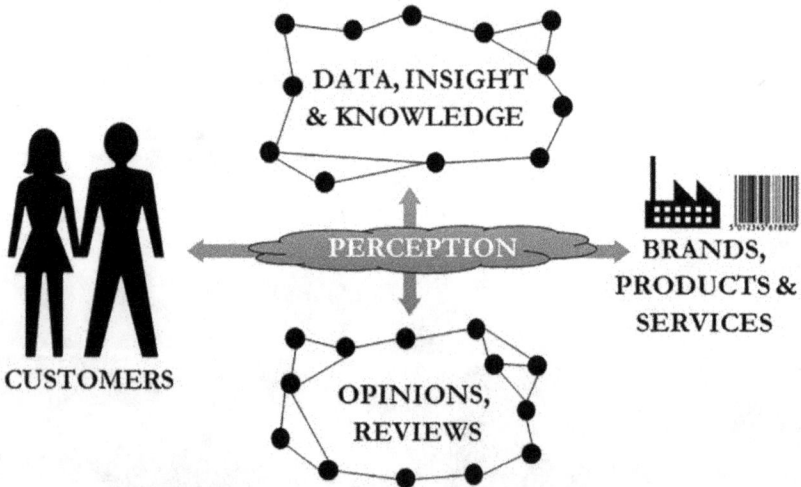

Customer perception is in the centre of any experience, and it is influenced by networks of inputs. It is almost like when mother-in-law sees a potential fiancé the first time and does an *"objective"* evaluation of the candidate based on her *"valid"* perceptions. Let's assume that in the picture above the mother-in-law is the customer and you, as a potential fiancé, are the product available. The mother-in-law, called Shirley, may have certain facts of you. Maybe she knows that you were born in the New York City. She might also know that you were in a foster home and got yourself in trouble often. You were expelled from several schools.

Those are the data, insights and knowledge she has available and it is affecting her perception of you. Then she might talk to her neighbour, called Harriet. She is well known in the neighbourhood for her observations (read: stalking). Harriet has formed an educated opinion based on her 12½ minutes of watching you coming and going from the house. She also has occasional notes available on your clothing and hairstyle. Harriet concludes you are a weird person (go figure!) Now Shirley is not only affected by the data she has available but also by the review given by her neighbour, too.

She will also have to factor in the direct experience she has had with you as well as her background: Shirley is from Denmark. She does not watch movies. She does not know that you are Sylvester Stallone[24]! And by this time, she is very … very … confused and opinionated! Luckily, she approves the transaction for a lawfully wedded partner No.3, and you can try to live happily ever after. What a mess! But such is customer experience with telecommunication companies, too.

The boring business version of the above story and picture goes something like this: various forces influence us as customers. And those forces that shape our perception of a brand, product or service need to be managed as part of the Customer Experience Management.

Now that we know the customer perception needs to be managed, one of the biggest perceptions in the telecoms industry to manage is TRUST. Various studies find trust to be the most important matter driving people to digital service providers or away from them. And it makes sense, as we are highly dependent on our connections and devices. I mean, what can you do without an Internet connection? Ask anyone younger than 30 years old, and they will die just from the

---

[24] The story is fabricated, but the facts are more or less correct, https://www.biography.com/people/sylvester-stallone-9491745

thought of being offline for more than 10 minutes.

Trust is so important that Temkin Group has started to benchmark it across industries and companies[25]. Guess how many telecom operators are on the top 15 list? That's right, none. And how many on the bottom 15 list? Two. In the UK, both Vodafone and TalkTalk have made their way to the bottom of this list. And that may not come as a surprise thinking of the news you hear of Vodafone customer service and issues.

Industry-wise, TV/Internet Service Providers are near the bottom of the list. Also, Wireless Carriers are below the middle line. Supermarket chains seem to be doing well, so perhaps there is something for the telco companies to learn from them? Looking forward, here are latest trends the industry practitioners, vendors and commentators see impacting the CEM[26]:

- **Digitalising customer experiences**
- **Customer loyalty and retention**
- **Data and analytics, contributing to customer insights**
- **Employee engagement**

The above points are all important forces shaping the Customer Experience Management efforts. As Communication Service Providers (CSP) are turning into Digital Service Providers (DSP), customer experiences move more and more into the digitalised. Such development is good for the customers as it frees them from the place, time and traditional channels that are suboptimal. No more queuing up in a store, but instant transactions online, anytime, anywhere. But it doesn't come without a cost. Here are the top 3 challenges the same three groups say we will face:

- **Building a sustainable, customer-centric culture**
- **Creating actionable insights from data**
- **Competing priorities**

If you have any CEM activities going on in your organisation, how much do the above challenges impact your efforts? I bet hugely. As we have discussed, CEM is a comprehensive growth strategy, and

---

[25] 2017 Temkin Trust Ratings, https://temkingroup.com/

[26] CX Network, The Global State Of Customer Experience 2017

that makes it harder to master. No firm can control all of the factors that impact customer experiences. However, there are three essential components all companies have to learn:

- **The Environment: This is what the customer will experience. The Environment includes your brand, products and services. It also consists of the channels and content you deliver. Your environment contains thousands of assets to manage.**
- **The Context: This is when, where and how your customer will interact with your Environment. It includes all the journeys, touchpoints and interactions.**
- **The Effect: The previous two points are the cause, and this is the effect. When you put all possible variations of your Environment together with the Context, the customer has gazillions of ways to experience what you have to offer. This explains why the same customer may be happy one day and unsatisfied the next, whilst going through a seemingly similar experience.**

The above explanation of difficultness of CEM is a little bit academic, sorry for that (I am a PhD after all). What we can learn from it is that since we simply cannot control every single experience our organisation goes through with its customers, we need to have robust systems, processes and tools in place that will maximise our likelihood for success. You are always betting against the odds, and things will go wrong! The trust is broken on those moments, so you need to be prepared and to proactively try to prevent them from happening, and then reactively fix them as well as possible when they do happen. Also, you won't be overwhelmed if you focus on the CEM as a System.

IT IS PRETTY CLEAR NOW THAT WE ARE IN THE CUSTOMER PERCEPTION AND TRUST BUSINESS. We are not in the business of just selling connectivity and digital products. So far, we have been looking into how CEM has an impact on us and how we can have an impact on CEM. To make the latter point more practical, here is an example list of focus areas that contribute to successful Customer Experience Management:

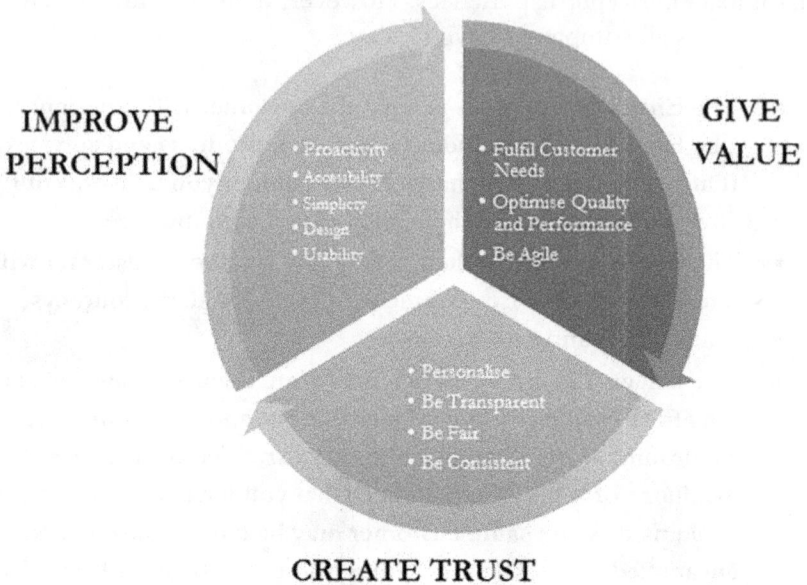

IMPROVE PERCEPTION

- Proactivity
- Accessibility
- Simplicity
- Design
- Usability

GIVE VALUE

- Fulfil Customer Needs
- Optimise Quality and Performance
- Be Agile

- Personalise
- Be Transparent
- Be Fair
- Be Consistent

CREATE TRUST

It is apparent that Customer Experience Management is not done in a vacuum. Why is it then that so many telco companies have set up a separate department for CEM and ask them to achieve their goals without company-wide collaboration? Because they don't know any better! A company managing their customer experiences in such a way will not be married to their customers happily ever after. So, take action today and work together!

CUSTOMER EXPERIENCE MATURITY - IS IT LIKE GETTING MARRIED TOO YOUNG?

"Customers are smarter than ever and looking for more value. More than just customer service, they want a great customer experience."
-Shep Hyken, Author, The Amazement Revolution

Teenage marriages are a great metaphor for Customer Experience Management Maturity (CEMM). They both pose difficult questions that are not only hard to answer, but will also depend highly on who we are talking about. Same way as some young people can make their marriages work long-term (my elder sister and her husband got together when they were 17 and are still happily married almost 30 years later) some companies can make their CEM efforts to work with very little maturity in the beginning. Then again, one of my acquaintances who is over 50 years old has been married six times since she was a teenager and has gone through extensive counselling to deal with the separations over the years, with no hope for a long-lasting relationship.

This resembles a company that tries to implement CEM and fails again and again (or fails once and then gives up) without learning from their mistakes. We can't say that just based on someone's age or maturity they will or won't be successful. S**t happens! But what we can say is that the maturity will impact the likelihood of success. And that's the best use of CEMM for us. We want to maximise our potential to be successful with our CEM program.

In general, maturity is a measurement of the ability of an organisation for continuous improvement in a particular discipline[27]. Maturity models imply that there is a certain predetermined optimal path for organisations to follow. As if one size fits all the

---

[27] Read more from https://en.wikipedia.org/wiki/Maturity_model

organisations! If you want to undertake a deep-dive into maturity and capability models, have a look at my book 'Business Process Management Capabilities'. In this book, we can have a lighter look on what kind of CEM Maturity models you should focus more than others.

As discussed earlier, Customer Experience Management suffers from vendor created confusion. Unfortunately, similar confusion applies to CEM Maturity. Too many consulting and technology companies are using CEMM as a way to differentiate their offering, thus creating new and scientifically unproven ways to measure and report on the CEM Maturity. And since there is currently a lack of globally accepted not-for-profit organisations that would own the standardisation for evaluating CEM Maturity, these vendors get away with what they do.

I am ashamed to admit that I have created four different CEMM models that are currently being used by reputable consulting companies around the world. Why did I do it? To differentiate those companies in the marketplace. Do they work? If used correctly, absolutely. They approach CEM Maturity from different angles, stressing different aspects of managing customer experiences.

But will they help you as a digital service provider to understand your current maturity level in a STANDARDISED way? Certainly not! You can still get value out of them, but you need to be aware that any vendor offering their CEMM model is ultimately doing it to benefit their sales pipeline. The biggest challenges these vendor-induced models have are:

- **Assumed linear advancement in the maturity scale. The assumption is that it can be determined on what level an organisation is on that scale. However, the real-world is much more complicated than that, and some aspect of the business may be on the highest level, some on the lowest. If this is reduced to a single average maturity level value, valuable information on the areas to improve can be missed.**
- **Most maturity models miss the need for the alignment between the areas it measures. As the maturity is reduced to a small number of categories, it is easily forgotten that those categories are not separate.**
- **The criteria for moving from one maturity level to another is**

often highly speculative and subjective. Who is then to say reliably what level the company should be on when individual aspects are on such different levels?

- They are all generic, assuming one size fits all organisations. That is rarely the case in reality.

- Most CEM Maturity models are Inside-Out, organisation-focused, rather than the customer (Outside-In).

Telecommunication companies have many different kinds of models for understanding their maturity in various business areas, like IT, processes and accounting. There are models like CMMI[28], eTOM[29], ITIL[30] and TL9000[31] that help telcos to standardise their operations. These have been helpful in the past, but to be honest, time is making them more and more obsolete. This is because the higher level you are on these standards, the more rigid the organisation becomes in responding to market and customer needs. They were fine when the business had time and resources for transformation programmes that took years. That is not the case anymore!

Thus, if there ever had been a standard for understanding CEM Maturity in the telecoms, it would have needed to be done in a way that is making companies more agile and proactive in meeting customer needs. And that is why I don't believe any of the vendors can create such a model. They are too focused on their solutions to truly add value that benefits everyone and not just their business (and sometimes their customers).

Thus, the solution will come from outside these companies, similarly as CXPA.org[32] has created the Certified Customer eXperience Professional (CCXP[33]) standardisation for the skills of individual CEM practitioners. Since we don't have something like

---

[28] Capability Maturity Model Integration
https://en.wikipedia.org/wiki/Capability_Maturity_Model_Integration

[29] Business Process Framework (eTOM)
https://en.wikipedia.org/wiki/Business_Process_Framework_(eTOM)

[30] Information Technology Infrastructure Library, https://en.wikipedia.org/wiki/ITIL

[31] TL 9000 is a quality management practice designed by the QuEST Forum in 1998, https://en.wikipedia.org/wiki/TL_9000

[32] Customer Experience Professionals Association, http://cxpa.org

[33] Certified Customer Experience Professional, http://ccxp.org

that available for organisations today[34] (and maybe it exists by the time you got your hands on this book), let's approach this challenge differently.

Let's first look into 5 main problems telecoms companies have, then go through existing CEMM models, and finally I will share with you a way to understand your current CEM Maturity without having to be restricted by any vendor-induced artificial levels.

Let's start with five key problems digital service providers will need to solve by increasing their capabilities in CEM (hence increasing their maturity):

- **Current revenue potential is decreasing as voice, SMS and data are commoditised and there are too many price wars taking out value from the industry.**
- **Over-the-top (OTT) players eat into traditional communication service providers' market share and revenues.**
- **Churn needs to be lowered and broadband de-commoditised.**
- **Networks get more complicated in every evolution (e.g. 5G), increasing network costs.**
- **The speed of telecom solution deployments has to move from months to hours.**

The cost of above problems is worth billions of dollars every year for the telecoms sector. They need to be solved as soon as possible. Unfortunately, not many telecom companies are considering CEM as a solution to these problems, though where CEM can truly help is with the first three problems: revenue, OTT and churn. By increasing the capabilities and maturity on Customer Experience Management, companies can increase their revenues, differentiate from OTTs (or partner with them) and reduce churn.

Let's summarise the kind of CEMM models we have available today, including most vendor created models, and we get a meta-model like this:

---

[34] TM Forum is not considered as they are too Inside-Out for us to talk about here

- **Strategic Value**
- **Value to Business**
- **Organisational Adoption**
- **Business Outcomes**
- **Things to Do**

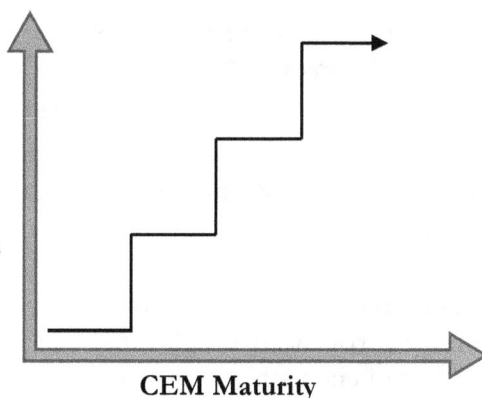

CEM Maturity

You can easily find many vendor specific, Inside-Out, focused CEMM models through an online search, if you are interested in the details of them. However, you have to understand that they are created with the main purpose of marketing and selling products and services, rather than genuinely offering the most suitable model for increasing your company's CEM Maturity (especially without their solutions). However, not all CEMM models are a waste of time. Here are examples of the benefits your organisation can gain through reaching higher CEM Maturity:

- **Provides senior leadership with an effective CEM roadmap to follow**
- **Sets out areas of responsibility and accountability across the organisation**
- **Mandatory in specific countries, if you want to tender for public sector work**
- **Communicates a positive message and intent to employees and customers**
- **Reduces your costs through LeanCEM**
- **Provides a model for continuous assessment and improvement**
- **Creates marketing opportunities (e.g. promoting reaching a new level in your maturity model)**

But, that's still so Inside-Out! We are more interested in how CEMM contributes to the lives of our customers! For instance, here are benefits CEMM has to offer for customers:

- Improved quality and service
- 'Right first time' -attitude
- Delivery on time and as promised
- Fewer returned products and complaints
- Independent audit demonstrates commitment to customer centricity

And to reach the above benefits for both, your company and customers, you need a right kind of CEMM model in place. You must have benefits bleeding out of your eyes by now, so, I promise that this is the last benefit list (in this chapter). To contrast the pain that you will get from poorly made vendor maturity models, these are the main gains from better CEMM models:

- Focuses on creating Successful Customer Outcomes
- Aligns the purpose of the business to customer strategy to business strategy to operations
- Specific to your business, market and context (no more one size fits all!)
- Gives a roadmap to plug the gaps and to improve continuously

Now that we know what kind of CEMM models are bad ones, how they can hinder your business and what benefits you can reach from a better model, let's end this thriller and see what a meta-model for such approach would look like:

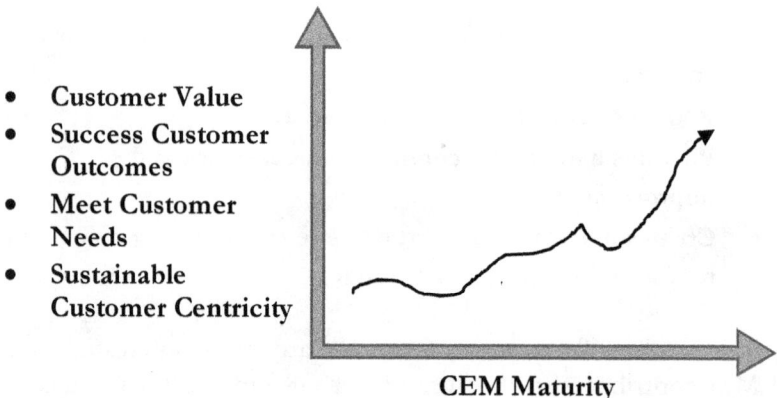

- Customer Value
- Success Customer Outcomes
- Meet Customer Needs
- Sustainable Customer Centricity

CEM Maturity

Since you made it so far, you can claim yourself to be the expert in CEM Maturity models in your business now! The above is enough information to have a profound discussion on what your maturity model should focus on and look like. You also are now knowledgeable in both the pain points and benefits available through such efforts. As the above-recommended model is a meta-model, and it may have a number of different kinds of implementations, let's look at an example of what a simple CEMM model derived from the above basis could look like:

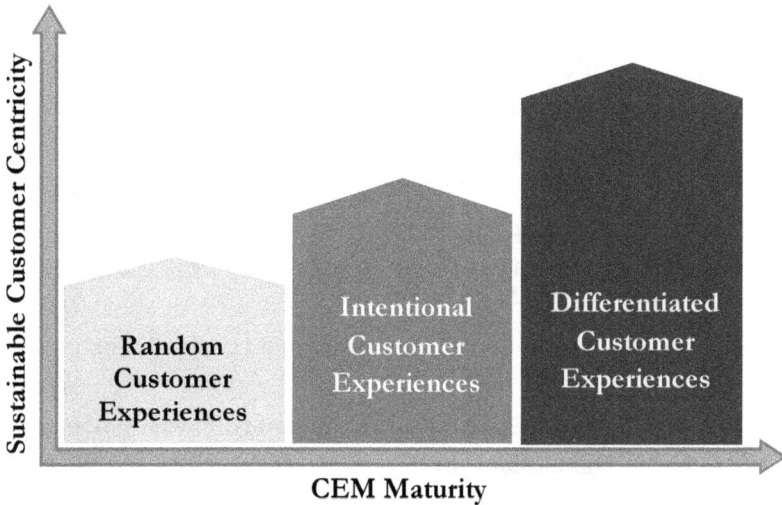

The above CEMM model may look simple, but don't let it fool you. It only has two levels to aim for as the first level is given. It simply suggests that you start creating intentional customer experiences and eliminate random ones. The problem with random customer experiences is that you cannot affect how the customer will feel about them. It leads to all kinds of experiences from downright horrible ones to most ecstatic. But who gets what kind of experience... only God knows!

Therefore, start by creating intentional customer experiences consistently. In the beginning, they don't even need to be amazing experiences as long as they are intentional. Once you start to move towards differentiated experiences, you start adding more great experiences (that are still produced consistently and intentionally). The 'Holy Grail' in the above model is to be able to produce great customer experiences consistently and to differentiate the customer experience with your company from the rest.

Customer Experience Management Maturity is about the progress of building a customer-centric culture into the business. It is a system of shared values and behaviours that focus employees' activity on improving the customer experience to gain, serve, and retain customers[35]. How will you know that you are on the right path then? Here are examples of leading indicators that you should see in your business (adapted from the previous footnote):

- **You have a functioning CEM governance model in place with true support from the C-level and the board.**
- **Everyone in the company can describe and are bought into your vision for intended and differentiated customer experiences**
- **Everyday jobs are aligned with your customer experience vision**
- **Performance is measured and improved from the customer perspective**
- **Successful customer outcomes become a standard, not a chance**
- **Customers are talked about in an enthusiastic way, not as a mandatory stakeholder that has to be kept happy**

Naturally, there are many more indicators that you may observe in your business as a sign of a 'customer spring'. Based on decades of experience, I know things are starting to move faster and in the right direction when you see people having excited discussions about how they have served customers better and how it shows in the business results. It is a great feeling, as you may know!

Since this book is about CEM in Telecoms and not just about the maturity, it is time to move on to other important matters. If you would like to try out a complimentary CEMM test, head down to the book resource page and sign up to the members only area[36]. You will find from there a survey form that will give you instant results, as well as a one-hour free consultation on your results from my team. That test is designed to get you started on CEMM quickly.

---

[35] Leading Indicators Of An Effective Culture Transformation by Samuel Stern, Forrester

[36] http://www.threecustomersecrets.com/member

HEAR IT FIRST-HAND: TELE2 RUSSIA

I love to learn more about new cultures and companies, how about you? Russia is famous for many things, but you wouldn't perhaps know that they also have a local telecom operator who has invested more in customer experiences than many companies in Western Europe. Tele2 was originally a Swedish based operator who entered the Russian market in 2003. However, since then the business has been acquired by a Russian investment company in 2013 and is now fully owned by Russians. They continue operating under the Tele2 commercial brand though. Today they serve over 40 million subscribers and have revenue over 2 billion USD. They operate mature 2G/3G/LTE mobile networks all over the vast Russia territories. They employ ~7,000 people.

I had the pleasure to interview Nina Gyubbenet, a Director of Customer Experience at Tele2 Russia for this book. She has been leading customer operations in Tele2 and other telecoms before that for over two decades. Nina's background is in customer service, and her roots are deeply in solving challenging customer problems with products, network, services and processes. In 2015, Tele2 Russia decided to focus strategically on Customer Experience Management. Currently, Nina's job is to build CEM as a system (should we call it CEMaaS?) into Tele2 and she has been doing it for the past two years rigorously.

When Nina started looking at the customer experiences Tele2 Russia provides, she had a very familiar problem for any customer experience professional, silo mentality[37]. She saw as the main challenge to solve the huge gap between what customers wanted from the business and how the business delivered it: "*Customers want a seamless experience. Also, silo mentality causes lots of inefficiencies and issues for customers*", she said.

That is why Tele2 Russia wanted to start a customer experience programme, to see what was happening in the business from the customer perspective. Nina shared that "*company's functional KPIs created a conflict between short-term business targets (such as revenue and profits)*

---

[37] "The definition of Silo Mentality is a mindset that occurs in organisations, which is inward looking and resists sharing information and resources with other people or departments within the organisation." (http://www.perceptiondynamics.info/silo-mentality/how-to-remove-silo-mentality)

and long-term customer loyalty (e.g. NPS)".

In Tele2 Russia, CEM started with a 360-degree customer view. When you transform that view into actionable insights, you will get powerful ideas on what improvements you need to focus on to make customers happier and the company more profitable. Customer Experience Management enables the company to generate sustainable revenue from loyal clients. In that scenario, tools like NPS are a means to an end rather than the end: *"NPS alone doesn't get enough attraction. Revenue and churn reduction speak much louder in the business. That is why connecting NPS to revenues is very important. CEM team has to sell their ideas through the lens of rest of the business to gain bigger attraction"*, **Nina** elaborates.

Today Tele2 operates through a sophisticated CEM framework that consists of four key areas with customer analytics in the centre:

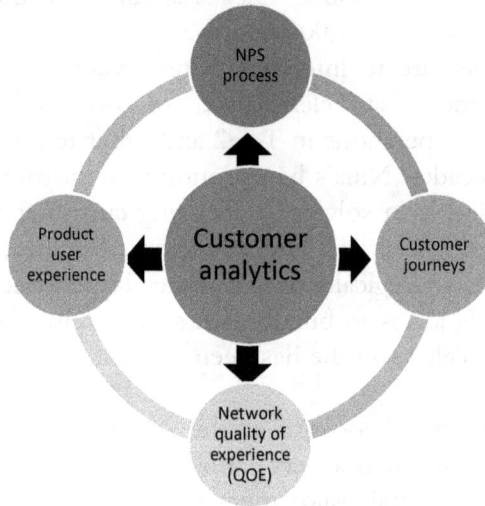

The customer analytics focuses on understanding individual customers. It is used to recognise both potential customer issues and opportunities. It enables Tele2 to understand root causes for churn and to identify any unsustainable revenues the business may be asking from the customer. The NPS process focuses on continuous improvements around customer loyalty and advocacy. One very important thing is that NPS became a bonus indicator for all management level employees started from C-Level. **There are 3 main outcomes from the NPS programme:**

- **Recognition of the importance of customer prospective**
- **NPS is measurable and manageable**
- **Dignificant increase of employee involvement in CEM initiatives**

The product user experience focuses on improving tariff plans, services, content and testing use cases. It doesn't focus only on usability, but the end-to-end customer experience. The customer journey analyses existing and desired journeys to identify gaps and opportunities to WOW the customer.

Finally, the network quality of experience (QoE) focuses on optimising capital expenditure (CAPEX) for network rollouts and capacity management. It tests and implements the modern concept of Network CEM – monitoring and managing customer experience per user per service. Such CEM framework has served Tele2 well in the past couple of years as we will soon see.

The governance of CEM is very important for any business. In the beginning there were many discussions around CEM responsibility areas; should it have direct governance over the important touch points like customer service, CBM, customer communication, etc. However, it was decided that CEM should not take over or substitute the existing functional roles.

In Tele 2 Russia, the CEM team is a transformation catalyst, trying a new customer centred approach in different areas, proving positive business impact. For example, CEM started Customer Journey mapping and product UX testing applying design thinking methodology – areas never believed and never touched before. CEM initiated massive shift from network KPIs to QoE KPIs, introducing to the company new network experience monitoring tools and step by step establishing new business processes.

The purpose of the Tele2 CEM team is to make other teams more efficient in using 360-degrees customer knowledge and change their business processes to a more customer centred approach. This is why **the CEM team in Tele2 operates with cross-functional collaboration across the whole business.**

The most interesting thing that Tele2 Russia has done is about "eating own dog food". **They didn't want to be one of those companies who are like councillors who give marriage advice while having been single all their life.** Thus, Tele 2 changed their corporate policy on employee telecommunication service compensation.

In many telecom companies, employees get their communication services for free or discounted from the company. Nina describes the problem it causes: "*They don't use the same processes as customers. They don't know the roaming costs, tariffs, or what services are available*", and continues, "*What we have done, is taken this benefit in this traditional form out from the employees and put them into the standard commercial tariffs. Now, employees have to follow the same processes as customers*". The company will then later reimburse the expense at the end of the journey, based on the level of the person's role in the company. **How amazing is that?**

Now that all Tele2 Russia employees are on a commercial tariff, they had to figure out what they pay for, how to pay for it, what services to have, and so on. They have to connect and disconnect services and to behave like a normal customer. To take the learnings from all this, Tele 2 has a company-wide digital portal for Tele2 employees where they can report CX related problems – Walk-The-Talk. The router sends the issue to the right employee who is able to solve it in a short time.

Walk-the-talk is not focused on customer service, but the wider customer experience, processes, communications, problems, and network issues, too. All employees participate this customer experience game and get points for CX issued discovery, which they can claim for rewards. As you can imagine, "*in the beginning the resistance was enormous as the employees came to see how difficult it is for the customers*", Nina says.

But they also appreciated to have this experience and to work on improving it for their customers and themselves. 70% of employees participate in this CX game. Such employees are hugely more invested in CEM initiatives compared to traditional companies who take advice from councillors who are not qualified (i.e. employees who do not use the services themselves).

Customer Experience Management, together with other business improvement initiatives, has paid a huge return on investment for Tele2 Russia. In 2017, they have been able to turn the long-term negative NPS trend to sustainable positive growth following with outstanding revenue increase far above the average market level.

Through this robust CEM programme, Tele2 has been able to identify that customer experiences have a significant impact on their bottom line. **Using robust customer analytics, they now know**

that 1-point increase in the Net Promoter Score equals to 0.5% increase in their revenue. And so we have learned more about this interesting telecommunication service provider in the exotic Russia. How can you apply some of these teachings to your company?

I. NAKED TRUTH: FINANCIAL IMPACT OF CUSTOMER EXPERIENCES

"Just one point increase on the Customer Experience Index would yield the telecommunication sector in the U.S. half a billion dollars higher
revenue per year!"
- Forrester Research, 2017

MAKE YOUR OWNERS HAPPY, PUPPY

This chapter feels a tad ironic from the Outside-In[38] perspective. Why do we spend so much time pleasing the people who own the businesses when the customers are the ones who can fire even those owners? I respect the resources, faith and commitment owners put in their companies, but we all need to start to acknowledge that they, too, work for the customer, and not for their wallets alone. In most businesses, such drastic change in the mindset of owners would lead from asking for short-sighted, unsustainable profits into building a customer-centric, healthy company that brings in the fruits for a long time to come.

How we can accelerate this change, is by taking the knowledge gained and the data presented in this book and educate the company owners and boards of the opportunities they are consistently missing. I suppose it's a bit like trying to break your father-in-law out of the decades-long habit of gambling on horses. We can do that by showing him statistics from the past decade on how much he has already lost and how much he will continue to lose if he keeps on thinking that it's the best "get-rich-quick-scheme" for him to invest in.

Gambling on horses may feel great temporarily on those Saturdays (or 'quartiles' in the business world) when he accidentally bets on a winning horses (or a company comes up with a product that sells well for a while). But, a financial short-coming is way more likely in the long run than success since it is not a sustainable strategy. In this analogy, the better option for the father-in-law would be to learn to trade with stocks (which is still gives the thrill of gambling but can yield returns more likely than horses). For companies, this would mean betting on their customers rather than Inside-Out product strategies.

---

[38] i.e. customer-centric

"You never change things by fighting the existing reality. To change something, build a new model that makes the existing model obsolete."
- Buckminster Fuller, American Architect

To promote a customer-centric mindset for business owners, **we need to demonstrate the impact of customer experience on the value of a company**. There is plenty of data available that shows this for various companies, both inside and outside of telecom sector. We will be discussing these companies and their results throughout the book in the case studies. But before we get into too many details, this is the main graph[39] from which you will need to convince your owners and the board about the business impact of customer centricity:

Total return percentage to owners of companies with above- and below-average ACSI scores over a 10-year period with a comparison to S&P 500 index.

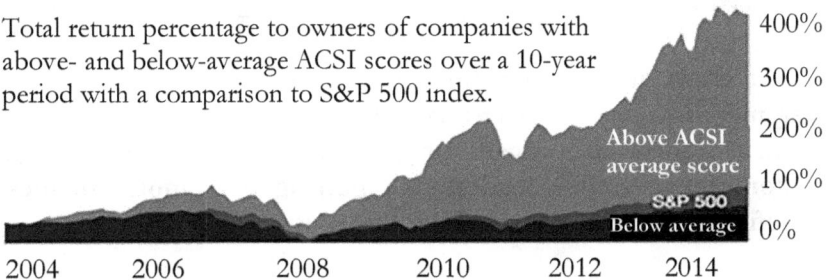

400%
300%
200%
Above ACSI average score 100%
S&P 500
Below average 0%

2004   2006   2008   2010   2012   2014

The graph shows how consistently **companies that are having above-average scores in the American Customer Satisfaction Index[40] (ACSI) are outperforming their competition, even compared to the Standard & Poor 500 (S&P 500) Index[41]**. Just looking at the year 2008, when the global economic downturn hit the world, those companies who were experiencing a high ACSI score were dropping in average only -28% while S&P 500 companies fell -42% and below-average companies fell even more.

---

[39] Putting customer experience at the heart of next-generation operating models by Shital Chheda, Ewan Duncan, and Stefan Roggenhofer

[40] The American Customer Satisfaction Index, http://www.theacsi.org/

[41] More information on S&P500 Index at https://en.wikipedia.org/wiki/S%26P_500_Index

Going back to our fictional father-in-law, you now have an investment strategy to suggest to him. By investing in companies that are customer experience leaders in a market, the value of stocks going up as well as the Return-on-Investment (ROI) in the form of dividends are much likelier than winning with betting on horses. If you are in the United Kingdom, you can use the UK Customer Satisfaction Index (UKCSI) or Temkin Group findings. For other countries, it is likely there is a benchmarking available locally.

An esteemed business publication, Harvard Business Review, undertook a study on how the performance of five business metrics compared to customer experience leaders against laggards. Their findings speak for themselves[42]:

BUSINESS METRIC	LEADERS	LAGGARDS	DELTA
Profitability	60%	35%	25
Customer Quality	66%	27%	39
Revenue	60%	28%	32
Market Share	54%	29%	25
Customer Retention	54%	20%	34

The above are percentages of companies receiving benefits from customer experience management. The delta means the difference between leaders and laggards in percentage points. **On average, customer experience leaders are getting 31% more business benefits from CEM across these five metrics than laggards.**

---

[42] "Lessons from the Leading Edge of Customer Experience Management", a report by Harvard Business Review Analytics Services, 2014

BRING HOME THE BACON

---

*"880 business leaders said their company realised
on average $3 in benefits for every $1 it spent on
improving customer experiences."*
- Avanade and Sitecore research, 2016[43]
(paraphrased from the research results)

By this time in this book, it is fairly evident that Customer Experience Management and positive financial results walk hand-in-hand, like a married couple on a romantic beach. Still, it is fairly common to come across the statement that the telecom companies are doing fine as they are. And that is true in a micro scale, looking at specific companies that are doing well. On a macro scale, the panic is rising! Communication service providers have relied on revenues coming in from traditional sources, such as data, voice and messaging.

OTT players have taken over these markets with applications that rely on the capabilities created by communication service providers while taking most of the revenues to themselves. This has a created a situation where traditional players need to find new ways to add value to their customers to plug the revenue gap that just keeps on growing. STL Partners had already made predictions back in 2013 that telecommunication players would have about 25% of a gap to fill in by 2020. By 2017, that gap has already grown to 30%[44]. Extrapolating from their data, we can predict a trend until 2025, and it is not looking good:

---

[43] Research by Avanade and Sitecore – Customer Experiences and Your Bottom Line, 2016, https://www.avanade.com/en/about-avanade/partnerships/sitecore/customer-experiences-whitepaper

[44] Sense check: Can data growth save telco revenues?, STL Partners, April 2017, https://stlpartners.com/research/sense-check-can-data-growth-save-telco-revenues/

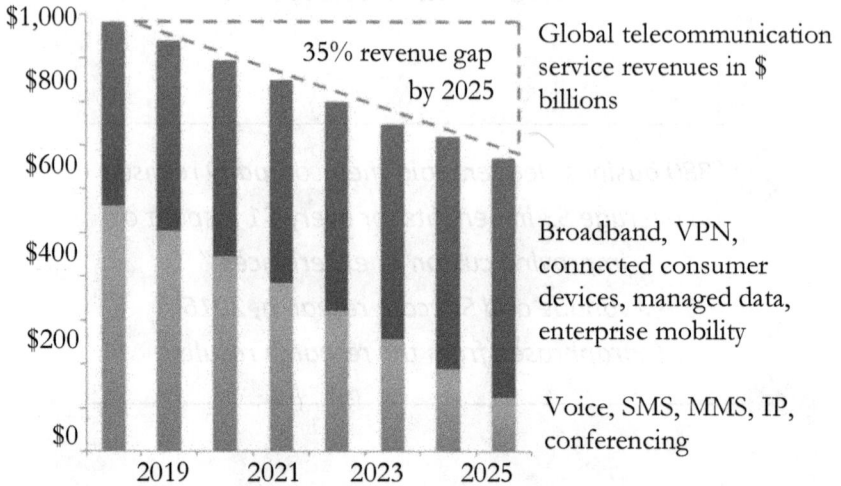

Extrapolating from the past does not always work as the industry may significantly change in the next five years and then this prediction wouldn't be accurate anymore. However, it does tell a story: if we don't change what we are doing in the telecommunication sector today, we are at risk of losing about $350 billion in revenue. That certainly keeps me awake at night!

As you may remember from the previous chapter, stocks of firms with higher ACSI scores have the propensity to do better than those with lower scores. This happens as satisfaction affects the willingness of customers to buy, and therefore affects revenue. Temkin Group conducted a study[45] on this matter amongst 10,000 wireless service clients and made these findings:

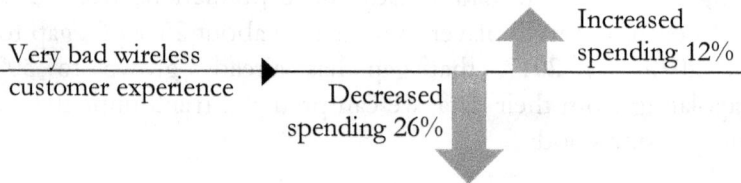

Clients reported that 13% of them had a very bad customer

---

[45] Temkin Group report, What consumers do after a good or bad experience, 2016

experience with their wireless service provider in the previous six months. The study looked into how their spending had changed since. 12% of those people had increased their spending while 26% had decreased their spending. This means a net sum of a -14% decrease in spending amongst those customers who had a very bad experience in the last half year.

Translating that to lost revenue, if there are 82 million wireless customers in the U.S. and an average revenue from those customers is 40$, then 82m * $40 * 0.14 = **$459m revenue has been left on the table because of those very bad customer experiences alone!** One's loss is another one's gain. That money has not disappeared from the market; it has just gone into someone else's pocket. And that's what we will look into next.

Avanade and Sitecore conducted a research back in 2016, asking company leaders about what kind of return they have received from their customer experience initiatives. **Close to 900 business leaders said their company realised on average $3 in benefits for every $1 it spent on improving customer experiences!** The most significant advantages they saw were around increased revenue (40%), better financial performance than the competition (38%), and improved sales cycles (37%).

On average, their performance increase in these three areas was around 20%. As you know, that is a very impressive return. But the benefits did not stop there. The respondents reported that better CEM produced heightened levels of customer satisfaction (58%), customer loyalty (45%) and better customer acquisition (41%). These areas too saw an average improvement of 19% to 22% in each area.

If you have a Voice of Customer (VOC) programme in place in your company, you can conduct your own analysis to see what exactly the linkage between customer experience and revenue is. That can be achieved by bringing in the value of customers and combining that data with customer perceptions. You can then tag those customers and see if their spending goes up or down and whether they churn or take up new products. You won't be disappointed with the insights acquired from such study! Meanwhile, we can take a look at the industry utilising data gathered from close to 500 companies in the U.S. and the UK by McKinsey Analysis and Temkin Research[46].

---

[46] McKinsey Analysis, 2015, Higher satisfaction leads to higher revenue growth & Temkin Research, 2016, Customer Experience Correlated to Loyalty

Customer experience &
customer satisfaction

Correlation from 0.78 to 0.89

Likelihood to repurchase &
revenue growth

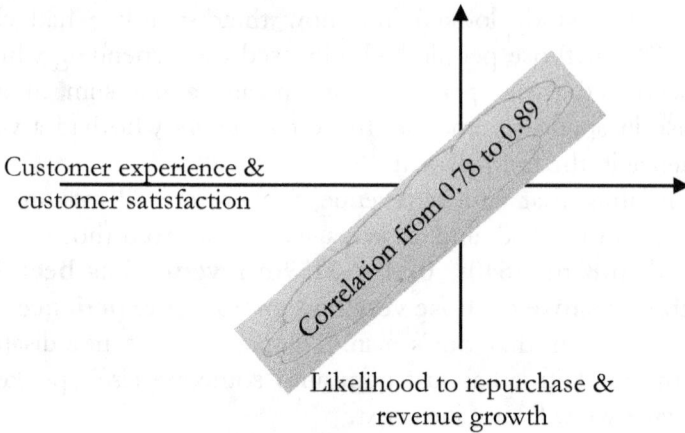

I try not to get too technical here, but the graph above puts together two datasets to save you from going throw several similar graphs. It shows how customer experience explains from 78% to 89% of likelihood to repurchase. As this data is based on an average of close to 500 companies, your numbers may look different. And yet the message is clear: **Customer Experience Management has a direct impact on revenue!**

Would you like to know how much bacon (or lettuce for vegans) you could bring home? That is possible. We have already looked through several numbers that give you an indication of the potential. But we can add more science to such evaluations for your business specifically. Feel free to insert your own numbers to below calculation[47] and see what you come up with:

---

[47] Adapted from Customer Experience 3.0 by John A. Goodman

```
                              50% satisfied      20% won't
                              with resolution    repurchase   = 1,250
                        5%
                   complain→  30% neutral        50% won't
Customers                     with resolution    repurchase   = 1,875
impacted by
CEX issues:                   20% unsatisfied    80% won't
250,000                       with resolution    repurchase   = 2,000
                   95% won't                     80% won't
                   complain                      repurchase   = 190,000
```

Total of customers at risk:	195,125
Average revenue per customer:	$40
Total revenue at risk due to these issues:	**$7.8m**

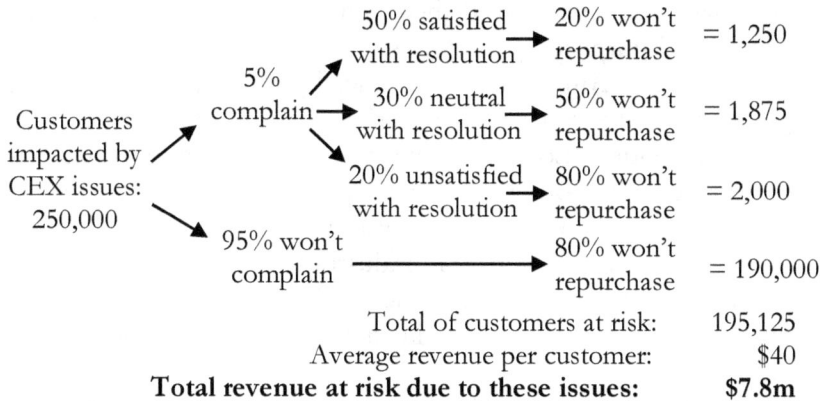

As you know, there are many ways to skin a cat. The above calculation is based on knowing how many customers are having very bad customer experiences and what their average revenue is. The other way to look at lost revenue is through a Voice of Customer key metric, such as Net Promoter Score or Customer Satisfaction. In that case, your calculations could look something like this:

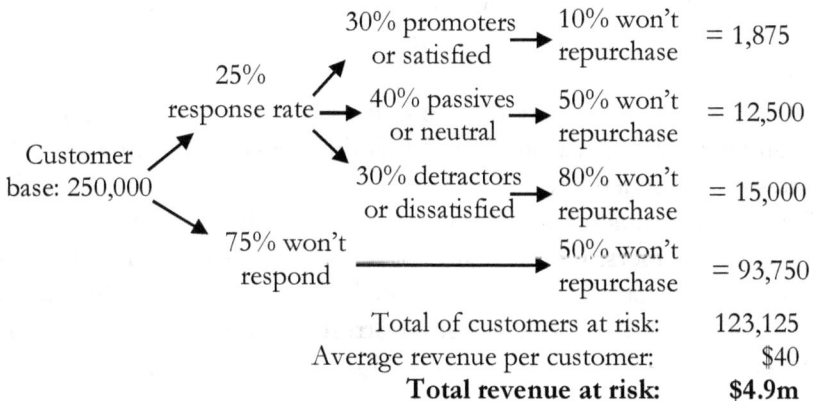

```
                              30% promoters      10% won't
                              or satisfied       repurchase   = 1,875
                        25%
                   response rate→ 40% passives   50% won't
Customer                       or neutral        repurchase   = 12,500
base: 250,000
                              30% detractors     80% won't
                              or dissatisfied    repurchase   = 15,000
                   75% won't                     50% won't
                   respond                       repurchase   = 93,750
```

Total of customers at risk:	123,125
Average revenue per customer:	$40
Total revenue at risk:	**$4.9m**

You have now both theory and practice available to discuss with your senior leadership. Let's go through a couple of case examples from the telecommunications industry to highlight the connection between customer experiences and revenue. The following cable company comparison was presented in a Forrester Research report[48]

---

[48] Customer Experience Drives Revenue Growth, Business Case: The Customer Experience

and has been referenced here under the fair use[49] policy for educational purposes.

In 2015, an American cable service by AT&T, called U-verse, was placed near the top of Forrester's CX Index ranking for internet service providers. Looking at the rankings, although U-verse was just as average as any other TV service provider, it was still significantly ahead of its competitor, Comcast, which was the lowest-scoring brand in both TV and ISP industries. The American Customer Satisfaction Index confirmed these relative rankings between these two companies. AT&T came in first as an ISP and second as a TV service provider.

On the other hand, Comcast was the last one as an ISP and second to last as a TV service provider. Forrester says analysing and comparing the performance of these companies was challenging as the financial information for U-verse was hidden deep in the annual report of AT&T. Similarly, Comcast's cable operation is just one of five business segments in Comcast's yearly reports.

Nonetheless, Harley Manning and Dylan Czarnecki were able to tease out the revenue and subscriber numbers for both companies' cable offerings. They found that from the year 2010 all the way to 2015, AT&T showed superior revenue growth compared to Comcast. AT&T's compound annual growth rate (CAGR) for revenue was more than 29%.

During that same period, Comcast's revenue grew less than 5%. To understand the reason behind this difference in revenue growth, Forrester examined the customer growth of both providers. While AT&T had double-digit growth among both its internet and video service subscribers, Comcast had only single-digit growth among its internet subscribers and negative growth among its video subscribers. The graph below (adapted from Forrester's report) illustrates this further:

Ecosystem Playbook, by Harley Manning and Dylan Czarnecki, June 21, 2016, Forrester Research
[49] Fair use policy explained in Wikipedia: https://en.wikipedia.org/wiki/Fair_use

CAGR comparison for AT&T as a customer experience
leader vs. Comcast as a laggard, years 2010-2015

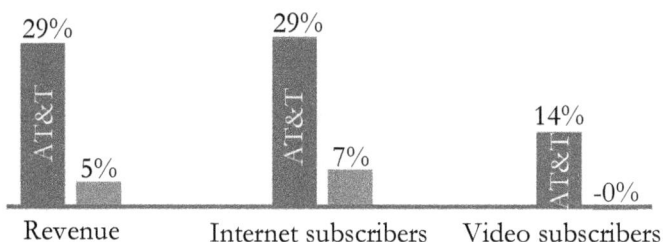

	Revenue	Internet subscribers	Video subscribers
AT&T	29%	29%	14%
Comcast	5%	7%	-0%

Our second case example for bringing home the bacon comes from the exotic Russia. Tele2 in Russia has invested in understanding their customer experiences deeply. They use close to 350 data points to understand what affects their customer experiences and Net Promoter Scores (NPS). **As part of that on-going analysis, they were able to uncover that every one percentage point change in their NPS score impacts their revenue by 0.5%[50].**

What about your company? Have you done the ever more important linkage between your Customer Experience Management metrics and business outcomes such as revenue? If not, the best time to start is today!

---

[50] Source: Interview with Director of Customer Experience at Tele2 Russia

FOOT THE BILL FOR CUSTOMER ACQUISITION

"My most brilliant achievement was my ability to be able to persuade my partner to marry me."
- Winston Churchill, Former British Prime Minister

All telecommunication companies are worried about their customer acquisition efforts. Either it is because they are losing customers and need to plug the gap or they want to grow and take market share from the competition. Net Promoter Score[51] has become of the most widely used customer metric in the industry. Funnily enough, it is mainly used as a customer experience or loyalty metric when I argue that its best use is for customer acquisition and retention.

Sure you can get insights in where the customer experiences are failing by analysing detractor data. And you can tackle customer retention through improving the experiences for both detractors and passives. However, where NPS is very powerful also is in mobilising promoters to undertake customer acquisition efforts on your behalf as well as maximising the lifetime value available from them.

Harvard Business Review claims in their article that **"depending on which study you believe, and what industry you're in, acquiring a new client is anywhere from 5 to 25 times more expensive than retaining an existing one"**[52] and that is very true. I have done this analysis in more than 20 companies, and if their data is reliable, the number has typically landed somewhere around 4-8 times. So, as the first step to foot the bill for customer acquisition, your business should look into customer retention (churn) and then in promoter activation to speak to their friends. There is telecoms

---

[51] More information at https://en.wikipedia.org/wiki/Net_Promoter

[52] The Value of Keeping the Right Customers by Amy Gallo, https://hbr.org/2014/10/the-value-of-keeping-the-right-customers

sector NPS data available to prove this point[53]:

Lifetime value of telecom customers times $1,000.
Categorised by the Net Promoter Score.

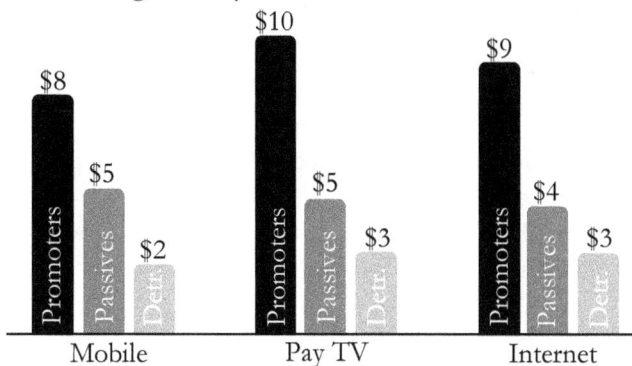

What is worth noticing in the above data is that moving a client from being a detractor to passive gives a much smaller return than moving them to promoters. There's a catch though. A client who scores 2 on a scale of 0 to 10 is very different to a customer who scores 6 (to become a passive) or 9 or 10 (to become a promoter). Such drastic changes are possible on an individual client level, but to produce such results on a bigger customer base will require providing much better customer experiences than the business is doing today.

Now we know that focusing on the existing clients is good business, not only from the perspective of reducing churn, but also activating promoters to bring their family and friends in (we'll talk about suitable use cases for doing that later in this book). Next step is to make buying easier. Forrester conducted an interesting research[54] with over 3,500 consumers on what were the key factors when deciding to buy a telecommunication service. The findings are as following:

- **Excellent customer support 79%**
- **Easy buying experience 78%**
- **Easy to use and mobile friendly website 77%**
- **Customer ratings and reviews on the service 65%**

---

[53] Great customer experience: What's the secret? by STL Partners in June 2017

[54] A commissioned study conducted by Forrester Consulting on behalf of LogMeIn, April 2017

Are you surprised? The above findings show that telco customers place more value on customer experiences than we may think. Other things influence their decisions besides price and promotions when making their decisions on what services to buy from communication service providers.

Further research from Forrester shows that companies who can make meaningful, emotional connections with their customers beat their competitors by 85% in sales growth. We'll talk more about emotional customer experiences soon. There is plenty positive feedback available from companies who have already engaged in Customer Experience Management activities. Here is a couple[55]:

Farmers Insurance   **Airbnb**

+30 on NPS
+3% retention

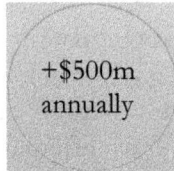 +$500m annually

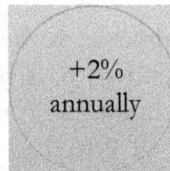 +2% annually

+Promoters refer friends 44% more often than detractors

How much revenue are you leaving on the table due to suboptimal customer retention and acquisition opportunities?

[55] Source: Medallia webinar that shared results from the Medallia Experience Conference, 2017

OH, IS S/HE GOING TO LEAVE YOU?

"Every contact we have with a customer influences whether or not they'll come back. We have to be great every time, or we'll lose them."
- Kevin Stirtz, Author, More Loyal Customers

Statistics are a funny old thing as they can hugely inform or misinform us. I am sure you have seen the news that half of marriages end in divorce. ABC and Fox News have been elaborate in their sharing of this faulty gospel. The Bravo TV channel went as far as introducing a divorce reality show, called *"Untying the Knot"*. An executive at the network called it *"a way to look at a situation that 50% of married couples, unfortunately, end up in."* Well, that's bollocks! New York Times debunked that myth[56].

As an example, only 30% of marriages that began in the 1990s did not reach their 15th anniversary. Those who married in the 2000s are so far divorcing at even lower rates. A similar type of skewed perception sometimes applies to telecommunication sector's customer churn rate[57]. As with marriages, there is always churn on telecoms, too. That is because we haven't so far figured an antidote to ageing and death (although I as a Christian believe there's sequel available after this premiere).

So, naturally, some of our customers die; according to statistical estimations, that counts about 1 to 2% of our customers. Analysys Mason estimated that in certain markets in Asia the churn us around 4% to 9%[58]. According to Marketing at Work[59], the average business

---

[56] Read the whole story at https://www.nytimes.com/2014/12/02/upshot/the-divorce-surge-is-over-but-the-myth-lives-on.html

[57] More information on churn rate is available at https://en.wikipedia.org/wiki/Churn_rate

[58] Research Survey Report, Connected Consumer Survey 2016: Mobile Churn And Customer Satisfaction In Emerging Asia–Pacific, Stephen Sale And Aris Xylouris

loses about 10 to 25 percent of its customer base per year. And Database Marketing Institute gives us these numbers[60]:

- **Wireless companies today measure voluntary churn by a monthly figure, such as 1.9 percent or 2.1 percent**
- **Annual churn rates for telecommunications companies average between 10 percent and 67 percent**
- **In the U.S. roughly 75 percent of the 17 to 20 million subscribers signing up with a new wireless carrier every year are coming from another wireless provider and hence are already churners**

I am fully confident that you already know your churn number. Therefore, the point is that churn is a real issue in the telecommunication sector and as we already discussed, worthy of solving with better Customer Experience Management. Every lost customer is revenue left on the table for the competition to pick up. I hope you are with me when I say that I would like to know how much customer experiences actually impact churn rates. Luckily, TM Forum has researched this in 2017, and the results look like this:

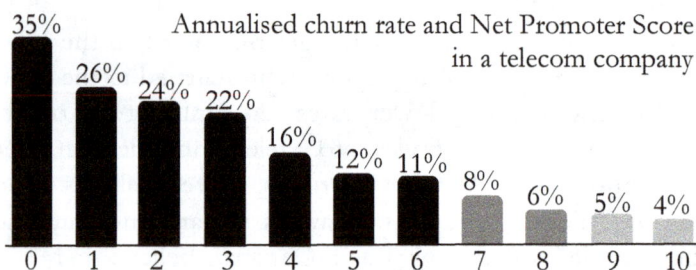

Annualised churn rate and Net Promoter Score in a telecom company

It is very easy to see from the above chart that happy customers simply churn less. Let's imagine that you were able to move your most unsatisfied customers (who scored 0) to be still unsatisfied but on the higher end of that scale (who scored 6). That would drop your churn rate for those customers by 24%. Depending on how many

---

[59] http://smallbusiness.chron.com/percentage-customers-business-lose-year-79271.html
[60] Churn reduction in the telecom industry By Arthur Middleton Hughes

customers you have in this category that could easily be worth millions of bucks!

You can modify the revenue calculation model we used earlier in this book to make your calculations. Did you notice something peculiar about the NPS with the above data? It reveals one of the biggest weaknesses in the Net Promoter Score methodology! Wow! Do you remember how the NPS is calculated? To refresh your memory here's the official equation[61]:

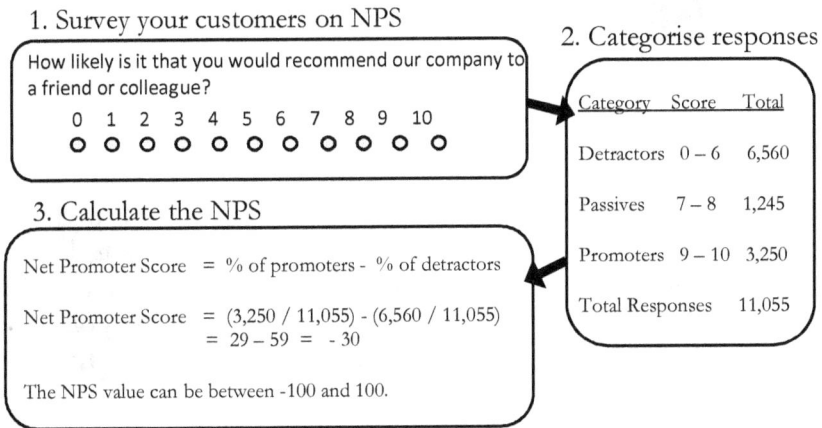

1. Survey your customers on NPS

How likely is it that you would recommend our company to a friend or colleague?

0 1 2 3 4 5 6 7 8 9 10
○ ○ ○ ○ ○ ○ ○ ○ ○ ○ ○

2. Categorise responses

Category	Score	Total
Detractors	0 – 6	6,560
Passives	7 – 8	1,245
Promoters	9 – 10	3,250
Total Responses		11,055

3. Calculate the NPS

Net Promoter Score = % of promoters - % of detractors

Net Promoter Score = (3,250 / 11,055) - (6,560 / 11,055)
= 29 – 59 = - 30

The NPS value can be between -100 and 100.

Using the above calculation as an only data point, we would completely miss the fact that customers who rate us as 0 (being detractors) are very different customers to those who rate us as 6 and are still detractors! The NPS is measured on an 11-point scale (0 to 10) but reported on a 3-point scale (detractors, passives and promoters) which loses a huge amount of granular data that would be beneficial for us when analysing churn rates.

This is a huge flaw in the standard NPS methodology, but it's a good thing we now know better and can take this into account when doing our NPS analyses. Well actually, thinking about the divorce rates again, is that really so? Let's see. We can use another research finding from the STL Partner's research referred earlier when we talked about the customer lifetime value. The below chart shows the churn rates for promoters, passives and detractors from their study:

---

[61] Adapted from http://www.davemitz.com/2011/04/16/3-easy-steps-to-calculating-net-promoter-score/

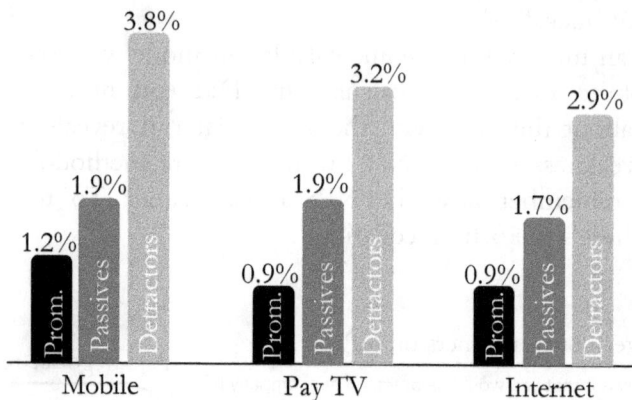

Phew, we don't have to divorce with Net Promoter Score just yet! Even categorised NPS can show the impact of customer experiences on churn. And we have learned that there are many ways to look at our CEM data about key business metrics, such as churn rates. You can be as elaborate in your analyses as needed to land the point that the **customer is your boss, no matter what level you work on in the company**!

As a recap, let's remind ourselves of why bad customer experiences are a detriment for your business. As we saw above, customer experiences and churn are linked. But we have to ask why that is. And the best source for that information are the customers themselves. Forrester put effort in the previously mentioned research to go out and ask from over 3,500 customers what they will do if they face a poor customer experience. Temkin Group[62] didn't want to be any smaller in numbers and went out to ask from over 10,000 consumers what will they do in case of a bad experience. This is what these two studies got back from respondents:

- **40% will tell my friends and family about the poor experience**
- **According to Forrester, 32% will stop doing business with that communications provider, while Temkin said, 18% stopped completely spending money with the company**
- **32% will immediately find an alternative digital service**

---

[62] Customer Experience Matters 2017 by Temkin Group

provider
- **21% of the customers decreased their spending**
- **11% will rant on social media**

If this starts to feel a bit repetitive at this point, that's great! Why? Because the telecommunications sector is still one of the most poorly performing sectors regarding customer experiences. The punishment for that is to see so much convincing data that you can't do anything else but to take dramatic action to improve the situation.

Now, let's move on to solutions. We will discuss use cases later in the book, but before that, it's good to explore what you can do about churn rates on a strategic-level. Understanding customer loyalty is a great place to start, and we'll talk about that in the next chapter. Meanwhile, let's play a little game. Here are typical actions companies take to manage their churn rates, have a look and choose the one your company currently uses:

ACTION	TASKS	BUSINESS MINDSET	CLIENT PERCEPTION
Level 0 – React	Improve retention rates and offers	"We need to stop losing customers."	"I wanted to cancel, but they made it hard."
Level 1 – Detect	Find potential churners and take action to prevent them leaving	"We need to prevent customers from leaving."	"I was about to leave, but then they fixed the issues I had."
Level 2 – Prevent	Proactively find and solve negative customer experiences	"We don't give customers reasons to leave."	"I have no active reasons to switch service provider."
Level 3 – Differentiate	Have a unique customer value proposition and deliver it exceptionally	"Add so much value that customers don't want to leave us."	"You should switch to my service provider as they are so great!"

What level is your company on? Now go and have so many pints, as you have earned it! If you were on level 0, you might want to watch this video *http://bit.ly/aol-cancelling* as that may be your destiny one day. If you are on level 3, then once you get rid of your hangover, you should go and share your story in industry conferences, as you are doing something great.

Medallia shared results from one of their telco customers, who is on level 3. That European service provider saw from their CEM data that promoters were 60% less likely to discontinue multiple lines of service (called cross-churn). Not only were they less likely to cross-churn, but they were also 110% more likely to take on new products with that multi-play operator.

DON'T LAYOUT MONEY ON OPERATIONS

"Only three things happen naturally in organisations: friction, confusion, and underperformance. Everything else requires leadership."
- Peter Drucker, Business Guru

Often, operational aspects of a business and customer experience management are seen as a separate thing. That is natural as most of the companies still function through traditional silo organisations (i.e. each are responsible for their departments where inter-departmental collaboration happens only when necessary). Thus, those people who work on IT and operators rarely think about the customers deeply. Typically, they talk about internal customers rather than paying customers. In the technology chapter of this book we will go through a practical approach on how we can align operators and information technology all the way to customer strategy, but before that, let's explore how operations and customer experiences could be a match made in heaven.

Operational expenditure (OPEX) usually refers to daily expenses for business functions, including wages, repairs, services, maintenance, utilities, advertising, supplies, insurance and legal fees. For the scope of our discussion in this book, we can use OPEX to refer to any costs that are incurred from serving customers. And the biggest sources of such costs in a telecommunication service provider's world are customer service (online and offline), network, and marketing.

We can leave the marketing out from this discussion as it is really up to the company to choose how much they want to invest in it. But customer service and network quality are mandatory features that affect customer experiences and operational expenditure. This is agreed by your peers, too. IQPC surveyed telecom customer experience professionals in 2017 as part of their CEM in Telecoms conference. The results show that when it comes to delivering long-term customer loyalty, customer service was agreed to be a key contributor by 48% of the respondents and network quality by 38%. To ensure we stay customer-centric, we need to ask the customers too. The following portion of customers indicated that these are key matters for them to stay as a customer with digital service providers:

- **81% indicated that having the best customer support tools available is important for them**
- **80% said that service providers need to offer customers support in multiple ways they like**
- **75% want service providers to contact the customers to avoid service issues from happening proactively**

Thus, customer service and network quality are extremely important and also very expensive. If you ask operational people what they are investing in, they will say into processes and information technology. If you ask that from CEM people, they will say customer journeys, data and analytics. What we need to do, is to pull these people together, to work towards a common goal. And that goal should be what CEM is all about, the sustainable and customer-centric growth of the business. While we'll talk about this alignment later in the book, **the concept of LeanCEM is extremely important here. It leads in to systematically streamlining both internal and external processes and fostering a culture of customer-centric continuous improvement.**

Most established telecommunication service providers are using Lean Six Sigma, TOM or ITIL. The biggest shortcoming of those methods is that the customer is the one something is done to, not the only reason for the business to exist. This is why **LeanCEM approach is much more effective** than those traditional ones. And how would such LeanCEM approach work?

LeanCEM MANIFESTO[63] by Dr Ohtonen

*Customers are our dear friends, to be helped **over**
being mere sources of revenue*

*We value sustainable goodwill revenue **over**
customer relationship deteriorating revenue*

*Our performance comes from value-adding
improvements **over** mere process optimisations*

*Innovation is the result of solving together with
customers **over** defined top-down company vision
and disciplined execution*

*We go to the customer to see facts and get
understanding of their world **over** making
assumptions and strategising in a vacuum*

*We encourage empathy, ingenuity, and passion **over**
compliance and exploitation*

*We empower employees to do the right thing both
for the customer and the business **over** instructing
every single process and decision*

*Challenge employees to better empathise customers
over just doing what management has told them to*

*No customer experience is ever perfect, and
continuous improvement is always possible **over**
thinking we are already good enough*

---

[63] I heard the word 'LeanCEM' originally from Tuukka Heinonen in 2014 in one of my CEM trainings. Today, I have defined what it means based on the Agile Manifesto, http://agilemanifesto.org

Do you think the above LeanCEM manifesto is asking for too much? Perhaps it is, but so was Agile Manifesto when it first came out and today only suicidal software businesses use the traditional Waterfall method it was directed against. In much the same way, not even in that distant future, companies that are not entirely focused on customers won't last in the market for long.

For that reason, companies that lay out money on inside-out operations will perform sub-optimally. What we need to do is to operationalise how we create customer loyalty and engagement. The chart below shows operational metrics for a customer-centric company that understands people from both emotional and behavioural perspectives:

	EMOTIONAL	BEHAVIOURAL
OPERATIONAL DATA	• Ratio of new customers to customers who left • Ratio of customers to employees • Quality performance	• Churn rates • Contract renewal rates • Usage metrics • Revenue metrics • Market share • Incremental performance
CUSTOMER PERCEPTION	• Customer satisfaction • Net Promoter Score • Likelihood to repurchase • First choice • Willing to forgive • Level of trust	• Likelihood to renew • Likelihood to churn • Likelihood to buy more • Likelihood to expand usage

The above listed ways to view operations, from business and customer perspectives, are on a very high-level, and it is not beneficial to go into too many details on this. What counts here is the thought. You can then apply it to whatever metrics you have in use today. Once you have done the above categorisation of operational metrics, you can start aligning them into CEM metrics. Here is an example how it could be done:

An example of how operational drivers may affect customer
experience metrics in telecoms

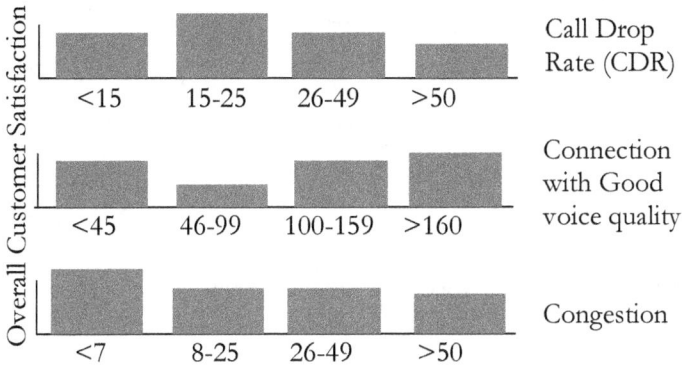

Instead of overall customer satisfaction, you could easily use Net
Promoter Score, Customer Effort Score or any other customer
metrics you have. In the same way, you can place whatever important
operational KPIs you have onto the right-hand side of the graph and
see how these two things affect each other.

Let's make things even more practical. AsiaInfo conducted a study
in 2015 to estimate how one specific customer experience use case,
Omni-channel, could potentially affect OPEX for the whole
telecommunication industry[64]. They evaluated that the value of an
Omni-channel approach from an OPEX perspective could be around
1.8%-4.8% in a typical western European operator. Here is an
example of building such a model:

---

[64] Quantifying the value of Omni-channel CRM for Telecoms by Katie Matthews, 2016

OPEX breakdown:	Improved CSAT → OPEX impact 0.3 – 1.0%	
34% Marketing and sales	Unified campaign channels → OPEX impact 0.3 – 0.7%	**Total potential OPEX saving 1.8% - 4.8%**
26% Network		
13% Customer service	Call Centre off-load → OPEX impact 0.9 – 1.4%	
5% Information Technology		
22% Other	Process automation → OPEX impact 0.3 – 1.6%	

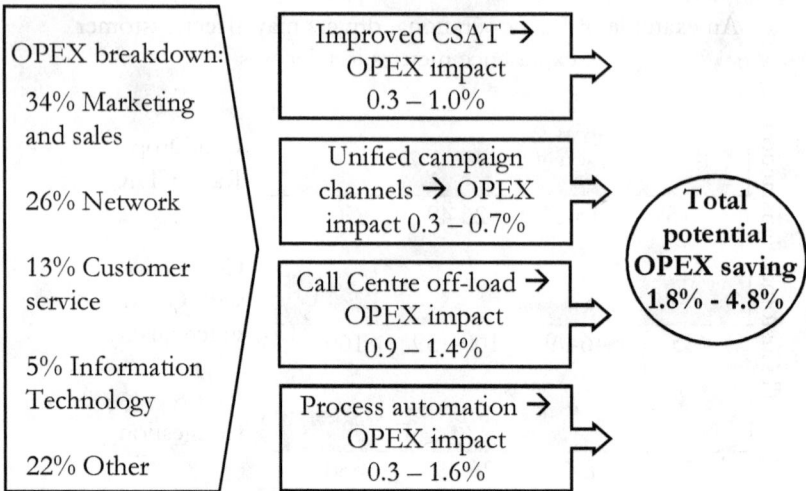

Though the above numbers would not directly apply to your organisation, you can re-use the model to link customer experience and operational KPIs and then estimate what kind of benefits are available for your business through specific CEM impacting-use cases.

In one of their webinars in 2017, Medallia shared results from such an exercise for an unnamed American retailer. The most remarkable aspect of those results was that just by having that company's employees to greet customers every time they enter the premises equalled to a 20% higher spend on that visit. If those sales associates went one step further and actually helped the customer on their visit, the spending by that customer climbed up by 30%.

The interesting fact here is that **our efforts to improve operations from customer experience perspective don't always need to be multi-million Omni-channel types of mega changes. Sometimes, simple changes in those processes that are most important for the clients will result not only in better customer satisfaction but also operational effectiveness and cost efficiency.** Where will you find a hidden gem in your company?

HEAR IT FIRST-HAND: TELUS CANADA

Telus telecommunication company is one of Canada's fastest-growing businesses, with over $13 billion of annual revenue and close to 13 million customers. Telus provides a wide variety of communications products and services, including wireless, data, Internet protocol, voice, television, entertainment and video.

In 2010, Telus made a strategic decision to focus on customer experiences as the primary differentiator from its competitors in the mobile market, which all offered similar products and services. Since then, Telus has focused on improving customer interactions to reduce the effort clients need to put into dealing with Telus. This is powered by a CEM strategy called 'Single Customer View' (SCV)[65]. To achieve this, Telus wanted to focus on these CEM improvement areas:

- **Engaging senior leadership and middle management**
- **Providing easy self-care for customers**
- **Recognising individual employee performance**
- **Removing silo mentality**
- **Integrating CRM, databases and channels for single customer view and Omni-channel experience**

It is well-known that the bigger a company is and more employees it has, the more likely that business is to have management that is removed from customers and their challenges. That is not something leaders typically go shout out from the rooftops, but it is an inevitable outcome of having so much to focus on in the business itself. In Telus, they decided several years ago that everyone in the company had to understand their role and contribution to customer value and experiences. This has led to a culture-shift in Telus, making the customer everyone's business. It has changed company policies, processes and mindset.

Providing easy self-care for customers has meant investing in tools

---

[65] "A Single customer view (SCV) is an aggregated, consistent and holistic representation of the data known by an organisation about its customers that can be viewed in one place, such as a single page or system." https://en.wikipedia.org/wiki/Single_customer_view

that enable customers to do that. As a consequence, Telus self-care apps in google and Apple AppStore have high ratings close to 5 stars. This is in high contrast to typical service provider apps that barely get in average 3 stars. They also offer one-to-one teaching to customers in their stores and have an active client forum with close 30,000 members.

The conventional industry problem of having reliable access to all customer data from multiple legacy information systems was a challenge for Telus, too. These systems are not designed for supporting personal engagement with customers. By using a state-of-art Customer Relationship Management (CRM) platform, Telus avoided the need to eliminate or replace legacy systems. They still use many legacy systems, but at least now all the relevant customer data is synchronised with the CRM system to ensure that employees have it available whenever needed.

It also ensures that all communication channels, whether manual or automated, have access to a consistent set of client data, and share a single customer view. As a concrete example of customer experience improvement, such an approach means that the customer does not have to tell their story multiple times to different Telus employees before their problem gets resolved, and the task is finished much faster and cost-effective.

To remove silo mentality and recognise employee performance, Telus has set up an internal, company-wide, social media network. That system allows its employees to monitor their performance, discuss with management, share materials, and give feedback. Creating a collaborative digital space has proved to be a significant motivator for employees to work together on shared issues. Since then Telus has launched various campaigns for employees to share their ideas and improve the employee and customer experiences. As an example, a front line ideation campaign generated over 3,000 ideas on how to improve customer experiences.

These investments Telus has made in their customer and employee experiences have not come without a significant return. Here are examples of the results[66]:

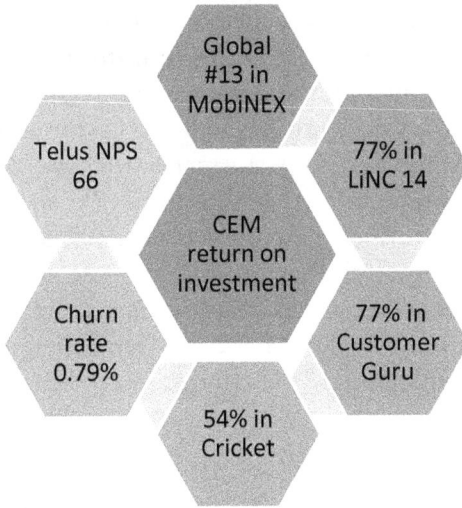

The exact scores above are not relevant as they vary depending on the source and time when measured. The main point is that Telus is performing better than any of its competition in Canada and the Cricket score it received was the highest score of any US mobile operator included. Also, the MobiNEX score places it in world rankings to position 13. Telus firmly believes that its investment in Customer Experience Management has led to its subscriber growth, churn reduction and high profitability. Here is a chart from Telus' Annual reports that shows this development from churn rate perspective:

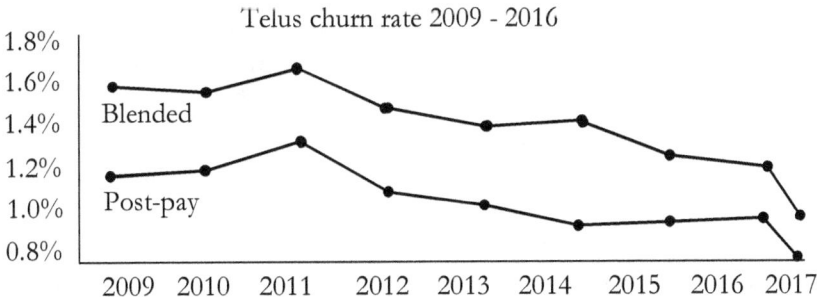

In 2017, Telus' consistent execution of Customer Experience Management has delivered an impressive wireless monthly post-paid churn rate of 0.79%, while blended churn was a 1.00%. Telus' post-paid churn rate has now been below 1.00% for 15 of the past 16 quarters[67]. **How are your churn rates looking in comparison?**

Next, let's look into the financial results Telus has received from its CEM efforts. As we already know, great customer experiences will translate into strong financial performance for telecom companies. In Telus' case, their key financial results of ARPU and EBITDA show steady improvement over the period from 2009 to 2017. While part of this success may reflect a general increase in the use of data services, their financial trend is clearly better than what competitors have achieved over the same period, when falling voice and SMS revenues have dominated ARPU figures.

Telus blended ARPU 2007 - 2016

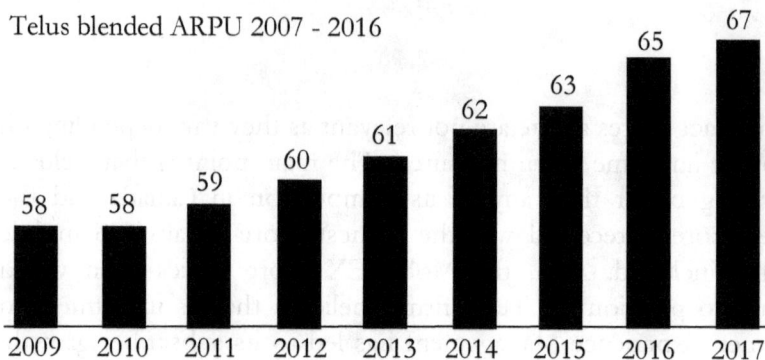

2009	2010	2011	2012	2013	2014	2015	2016	2017
58	58	59	60	61	62	63	65	67

Above is a very impressive Average Revenue per User trend from Telus! As long as they can keep up the good work on their CEM, the results should keep on coming. What has been common in both case studies, Tele 2 and Telus, is that both companies have taken Customer Experience Management seriously and found their unique ways of implementing it.

Also, return on that CEM investment in the form of higher revenue and smaller churn rates is a common denominator. Both organisations emphasise the importance of building a customer-centric culture and mindset through the organisation. Also, a certain level of humbleness is typical in these companies as they celebrate the results but also acknowledge that there is still plenty of work to do. **Customer Experience Management is a dynamic target!**

---

[67] http://www.marketwired.com/press-release/telus-reports-strong-results-for-second-quarter-2017-tsx-t-2229887.htm

II. CUSTOMER EXPERIENCE MANAGEMENT FRAMEWORK FOR BETTER AND FOR WORSE

"It is not the employer who pays the wages.
Employers only handle the money...
It is the customer who pays the wages."
- Henry Ford, The Inventor of Faster Horses

A CEM FRAMEWORK? WHAT'S THAT?

"Focusing on the customer makes a company more resilient."
- Jeff Bezos, CEO, Amazon

Customer Experience Management is both an exciting and a challenging business endeavour that holistically touches all areas of business. It is widely acknowledged that people engaged in leading CEM activities need to hold a vast variety of business, organisational and technological skills, much more comprehensive than most other leaders of a company. This is mainly because other C-suite leaders get to focus on their respective areas (contributing to silo mentality), while Chief Customer Officers (CCO) do not have any silo (narrowly defined business area) of their own. Instead, they need to collaborate across all departments of a company to change how that company operates to produce customer value and exceptional customer experiences[68]. Any company who sets their CCO to do this in a silo is set to fail.

What a CEM framework enables the company and the CCO to do is to establish a clear strategy and clear tactics for serving customers. The better the framework performs, the more cost efficiently the business can produce great customer experiences and value. In some ways, a CEM framework is like a marriage certificate and prenuptial agreement (pre-nup) together. It states the why, what and how of the business serving its customers.

It also describes how the relationship lifecycle ought to perform. As we know, it is easy to add brand promises on a company website, stating the wildest dreams the business might deliver to its clients. However, it's an entirely different matter to actually to deliver on

---

[68] Chief Customer Officer 2.0: How to Build Your Customer-Driven Growth Engine by Jeanne Bliss

those. For that, a CEM framework is a tool not to be overlooked by the senior leadership, nor shareholders. There is a catch though. No plan without execution is going to help a business to win customers. Thus, any reasonable Customer Experience Management framework includes a roadmap for setting the strategy in motion.

We have already touched upon the matter of vendor bias on many customer experience related issues, whether that is maturity models or CEM solutions. Unfortunately, this same trend applies to CEM frameworks. Most frameworks available today are created to solicit products and services. Consulting and solution companies are notorious for producing seemingly "best practices" that are sold to operators as the best next thing since the sliced bread. The caveat in such approaches is that customer experiences create a real competitive advantage, but only if the business can deliver it uniquely and genuinely. CEM is so dependent on the culture of a firm that it directly cannot be copied from one-size-fits-all frameworks and expect to work magically.

The CEM framework presented in this book is not supposed to be the best solution for any and all telecommunication companies. It is a description of necessary aspects to consider in a telecommunication sector specific Customer Experience Management framework. It can be used as a basis for developing further a CEM framework that works for your business correctly. As customer experiences touch all areas of business, it may make CEM frameworks heavy to comprehend and manage. They can consider hundreds of management areas that impact customer experiences, many of them overlapping with existing organisational and management practices. However, there's no need to re-invent the wheel here. We can utilise existing management practices (with realignment to CEM) to support CEM framework. For example, if your business already uses ITIL[69] or such, you can still keep using them, but just with added alignment to the customer.

The below CEM framework represents a comprehensive, end-to-end CEM framework that aligns all critical aspects of customer experience, business and technology management practices. We will discuss the alignment in much more detail in the next chapter and focus on the content of the framework in this one.

---

[69] More on ITIL at https://en.wikipedia.org/wiki/ITIL

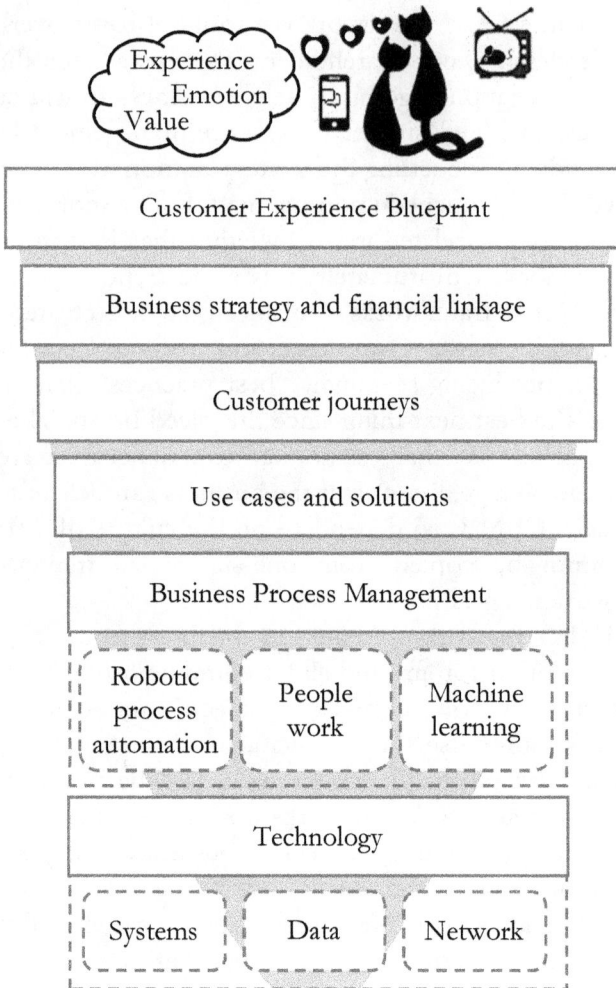

The above CEM framework is a simplified version of something that could potentially be overwhelming. What we want to do is to avoid complicated models such a TM Forum Frameworx[70]. These monstrous frameworks are very detailed, but also extremely difficult to comprehend and implement. And, most importantly, areas of them may or may not apply to your business.

The above CEM framework will apply to your business because it is a "minimum viable product" for CEM to perform well (yes, you can do with much less if suboptimal results are acceptable. You won't need this all at once either, as you can build it through your CEM

---

[70] https://www.tmforum.org/tm-forum-frameworx/

roadmap). A risk, if these aspects are not appropriately dealt with, is that the success for significant customer experience impact is unlikely. The above framework omits several important details that should be considered as the CEM maturity of an organisation rises:

- **Governance**
- **Customer service**
- **Leadership empowerment and support**
- **Capability management**
- **Proposition management**
- **Organisation, cross-functional collaboration**
- **Customer analytics and reporting**
- **Member-to-member communities**
- **Customer rewards**
- **Building brand equity and awareness**
- **Voice of Customer, Employee and Process programmes**
- **Customer metrics, such as Net Promoter Score, Customer Satisfaction, Customer Effort Score and others**
- **Customer complaints management**
- **Omni- and multi-channel communications**
- **Change, programme and project management**
- **Culture changing**
- **Employee and leadership coaching and training**
- **Employee recognition**
- **Employee journeys**
- **Employee experience blueprint**

We shall continue discussing this framework and specific elements of it through the rest of the book. Also, in the next chapter, we will look into aligning these layers of Customer Experience Management with each other for best results.

ARE WE GOING TO BE CUSTOMER-CENTRIC TOGETHER?

---

*"There is only one boss. The customer. And s/he can
fire everybody in the company from the
chairwo/man on down, simply by
spending his money somewhere else."*
- Sam Walton, Founder, Wal-Mart Stores Inc.

Working together with all stakeholders (customers, shareholders, leadership, management, employees, partners, etc.) is paramount for successful Customer Experience Management. There are very few businesses today, if any, where some form of internal or external collaboration is not needed. Typically, such cooperation is between internal departments (silos) of a company, though many times that is impaired by the silo mentality that shows itself in many forms:

- **Hierarchical silos – organisational-level based**
- **Operational silos – functionally based**
- **Channel silos – interaction based**
- **Geographical silos – location based**
- **Capability silos – competence based**
- **Product silos – proposition based**

Organisations need to foster key stakeholder collaboration and design customer (CEX), employee (EX), and partner experiences (PX) systematically to overcome the above silos:

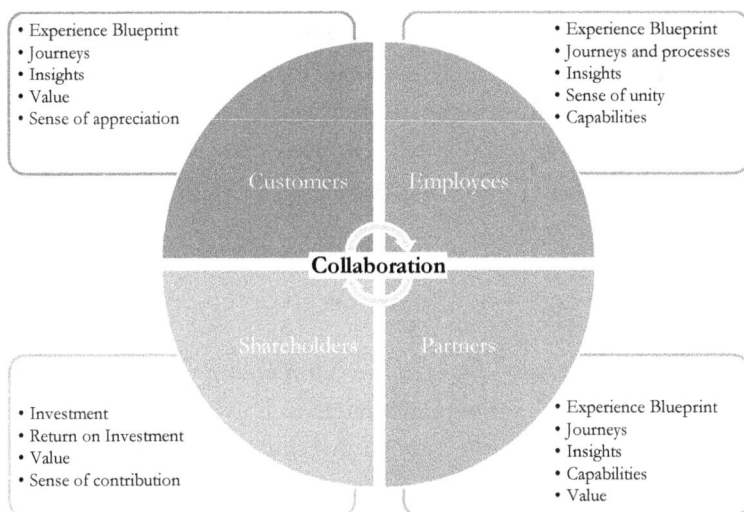

Customers
- Experience Blueprint
- Journeys
- Insights
- Value
- Sense of appreciation

Employees
- Experience Blueprint
- Journeys and processes
- Insights
- Sense of unity
- Capabilities

Collaboration

Shareholders
- Investment
- Return on Investment
- Value
- Sense of contribution

Partners
- Experience Blueprint
- Journeys
- Insights
- Capabilities
- Value

Mastering the experience for all relevant stakeholders will take the business to a new level. And it does not have to be anything super-complex, just thoughtful, as Jeff Weiner said:

"Inspire, empower, listen & appreciate.
Practicing any one of these can improve
[customer], employee, [and partner] engagement;
mastering all four can change the game."
- Jeff Weiner, CEO, LinkedIn Corporation
(stakeholders in brackets added by the author)

When you want to transform a company into a more customer-centric organisation, the most important factors to be considered are:

- **Proactive role modelling by the leadership and middle management**
- **Manage the employee experience so that people will want to be more customer-centric**
- **Always grow the organisation's ability to learn and change**

- **Set clear internal ownership and alignment for all experiences**
- **Focus all company efforts towards customer expectations and value by transforming customer wants and needs to internal objectives that are aligned and can be tracked**
- **Align the technology into the Customer Experience Blueprint**

There are many practical steps you can take to achieve the above and motivate employees to be more customer-centric:

- **Communication – Be clear on what customers expect from employees. Give practical advice on how employees can consider the customers as part of their daily work.**
- **Rewards and recognition – Give rewards to those employees who go the extra mile for customers. Recognise great performance.**
- **Training and mentoring – Train for skills and mentor for the experience. Run through customer scenarios and situations where employees can consider the customer in the best possible way.**
- **Integration – Remove silo mentality and empower employees to take responsible action to create great experiences.**

Let's take a practical example of changing a traditional view into a customer-centric perspective in telecommunications. Typically, communication service providers look at the customer lifecycle from their perspective (Inside-out) rather than customer's (Outside-in). These two worlds look very different:

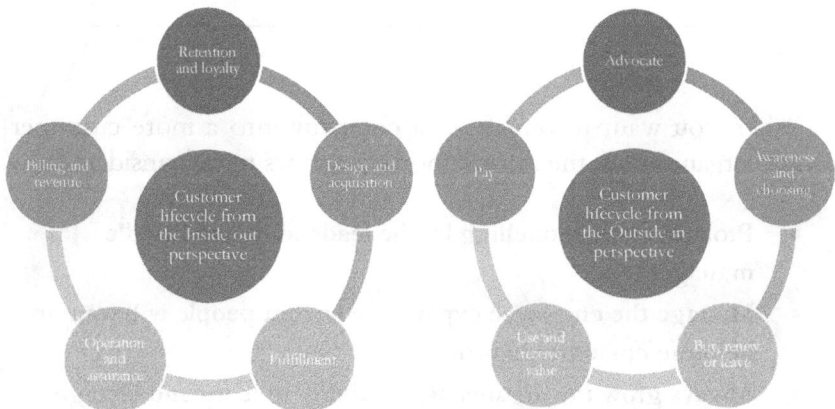

As you can see from above, a simple change in perspective leads to a different model (and therefore different outcomes). The focus shifts from internal activities to external results. Such focus shift can be achieved with all stakeholders as part of the experience blueprinting and journey mapping exercises.

Though this book is about customer experiences, there is strong evidence suggesting that employee and customer experiences are correlated. And even if they weren't, there's compelling evidence to not only focus on the customer experiences but also employee ones[71]:

- **Companies focusing on employee experiences (EX) were 4.5x more frequently listed on 'Most Innovative Companies' -lists by Fast Company, Boston Consulting and Forbes, than non-EX companies**
- **They are six times more likely listed on GlassDoor's 'Best Places to Work', Fortune's '100 Best Companies for Millennials' and LinkedIn's 'Most In-Demand Employers'**
- **Up to 3x more frequently listed on Brand Z and Forbes' 'Top Brand Value' companies**
- **40x more frequently in the list of companies whose impact is disproportionately large compared to other organisations**
- **2.1x Average Revenue vs non-EX companies**
- **4.4x Average Profit vs non-EX companies**
- **2.9x Average Revenue per employee vs non-EX companies**
- **4.3x Profitability vs non-EX companies**

The above list is desirable, isn't it? A shift from just CEX to include also EX is firmly rooted in an evolution taking place in organisations around the world as companies transition from treating people as assets or 'capital' to treating them as human beings. Such change, from forcing employees to submit to and work within specific processes and norms (i.e. managing assets), to understanding them and how they work best and designing solutions around their personal needs (i.e. employee experience). Jacob Morgan relates how EX is *"really about building that connection and relationship with your employees and, in a sense, giving them a type of ownership to shape the*

---

[71] Source: Jacob Morgan referenced in https://www.linkedin.com/pulse/how-build-ex-centric-organization-elliott-nelson/

organisation they are working for".

This chapter was about understanding how important focusing on other stakeholder experiences, beyond customer experiences, is beneficial for a firm. It does not only make Customer Experience Management easier in a telecom company but yields benefits of its own.

AND WHAT EXACTLY ARE WE SUPPOSED TO DO ANYWAY?

"A customer is the most important visitor on our premises; s/he is not dependent on us. We are dependent on him/her. S/he is not an interruption in our work. S/he is the purpose of it. S/he is not an outsider in our business. S/he is part of it. We are not doing him/her a favour by serving him/her. S/he is doing us a favour by giving us an opportunity to do so."
-Mahatma Gandhi, Civil Rights Leader, India

We want to fulfil the purpose of CEM, which is a customer-centric and sustainable growth for the business. To achieve that we don't only need a high performing CEM framework, we also need a plan on how to run it. As that is something every company needs to design for themselves, this chapter looks into reasons why to do it and then outlines a process for designing a customer-centric transformation for your business.

If you think about it in the context of marriages, it is one thing to say the wedding wow and another to keep it. If you don't have healthy coping mechanisms available for hard times, the divorce is inevitable. Much in the same way, if a company doesn't have practical ways to stay customer-centric, it won't last for long. We have seen so many companies who come out with great promises that are forgotten the first negative quarter hits the business (and they fall back to collecting short-sighted revenues).

Digital customer experience transformation is a way to change a traditional Inside-out organisation into Outside-in one. TM Forum asked in their 2017 survey, why communication service providers want to do that. Two out of three providers said stronger customer relationships is the main driver for their digital transformation aspirations. Also, Heavy Reading[72] conducted a study in 2017, looking into the benefits service providers have received from better understanding and alignment to customers. Their findings were the following:

- **More personalised offers increasing offer uptakes**
- **Customised self-service applications decreasing customer service costs**
- **Omni-channel customer service improving customer satisfaction**
- **Social media engagement and service driving higher advocacy**

Ernst Young[73] conducted a study looking into what telecommunication companies will mainly focus on in their efforts to drive customer-centricity in their business. The main reasons your competition is investing in Customer Experience management are:

- **Improving levels of customer service**
- **Creating more personalised customer experiences**
- **Improving network Quality of Experience**
- **Strike partnership with other industry players**

Now that we know what your peers are up to, we cannot avoid the question of defining the CEM aspirations for your business. Take time to answer the below questions and then have a mini-workshop with your colleagues to discuss further:

- **What is your company's appetite for change in the near, medium- and long-term?**

_____

_____

_____

_____

_____

---

[72] The state of digital transformations in telecommunications by James Crawshaw, Heady Reading
[73] Global telecommunications study: navigating the road to 2020 by Ernst Young

- Should your goal to be to change the customer experience fundamentally or simply to improve it at the margins? Why?

_____

_____

_____

_____

_____

- What are the gaps between needs, wants and expectations of your customers and what they actually experience with you?

_____

_____

_____

_____

_____

- How can the company gain a customer experience advantage against competitors?

_____

_____

_____

_____

_____

- At which point in the customer experience should the company concentrate to have a real impact?

_____

_____

_____

_____

_____

- **How do the overall capabilities of the employees support the customer experience the company wants to provide?**

_____

_____

_____

_____

_____

None of the questions above are easy to answer. And we are just getting started. Soon we will go through a Customer Experience Blueprint technique that digs even deeper into above questions. So you might as well answer the above questions as a warm-up. While you are at it, you might also want to think about the question of: how will others know that you are customer-centric? As this topic is not something that can be done in secrecy. So, what are the type of characteristics you would like people to use when they describe your new customer-centric business? Here are example desirable targets you might want to consider[74]:

- **Be ATTENTIVE – No customer requests or questions are left unhandled. The goal is to be seen as flexible and available. The customer should feel every effort is being made to accommodate their personal needs with a customised and relevant solution.**
- **Be PROACTIVE – The digital service provider anticipates customer concerns and proactively addresses them early before they become an issue. It is imperative to choose the right time, context and channel for interaction with the customer.**
- **Be CONSISTENT – For a great customer experience, the relationship with customers across channels needs to be seamless. The customer expects the communication service provider to have a record of all previous contacts so that they do not have to repeat themselves or receive conflicting**

[74] Source: Generating actionable insights from customer experience awareness by Jörg Niemöller, Nina Washington, and George Sarmonikas in 2016

information from different touchpoints. Information needs always to be consistent, flawless and immediate.

- Be ADAPTIVE – Services and products are continuously improved to meet customer needs, wants and expectations. The customer feels their needs are met and experiences any changes as improvements.

The above list speaks to what kind of business your company has to be. And we have already discussed why CEM is so important. However, we have not discussed one important aspect yet, that is what the value looks like from the customer's perspective. It is evident that we have to deliver value to the business, to shareholders, and to many other stakeholders. And for most stakeholders, that value shows itself in the form of financial matters (revenue, profit, cost efficiency, etc.) and potential (market share, products and services, etc.). But for customers value is so much more complicated. It is a network of personal, emotional, functional and other aspects that come together. As we already defining the characteristics of a customer-centric business, so let's also do an academic exercise of defining the characteristics of customer value[75]:

Social	Personal	Emotional	Functional
Self-transcendence	Gives hope	Gives happiness	Saves time
Social status	Self-actualisation	Reduces fear & anxiety	Makes money
	Motivates	Rewards	Reduces risk
	Belongingness	Attracts	Connects
		Trustworthiness	Quality
			Variety

---

[75] Adapted from: The elements of value, HBR, 2016, https://hbr.org/2016/09/the-elements-of-value

The above list is an example only. Well designed and performing Voice of Customer programmes can give you a similar kind of record for your customers. It would be so much simpler if I could just share facts with you, but the problem in averages and research is that they give generic (or academic) guidelines on the type of things you should look into, but every business is different and therefore the list of what your customers value the most is different.

Instead of copying & pasting details from this book, you should absorb the ideas and then implement them yourself. This way you can ensure that the most relevant answers are given to these questions. Speaking of which, we[76] have created a practical tool that you can use to create a Customer Experience Blueprint for your business. It is in use in hundreds of companies that work on increasing their customer-centricity. My previous book, *The 5-Star Customer Experience*, gives you detailed instructions on how to run a CEX blueprinting workshop. The following picture shows you a simple Customer Experience Blueprint canvas that you can use with your team to define a strategy for your customer experiences. The template is available for free download from the book support materials website:

---

[76] Tuukka Heinonen from Provad had kindly created the first version of the Customer Experience Blueprint canvas based on my previous book and I have further refined it for this book

Customer Experience Blueprint

Designed for:

Designed by:

On: dd/mm/yyyy

Iteration #

1. Target Customer
Customer archetype, role, motivation

2. Stakeholders
Key stakeholders and influencers

3. Goals
Customer Job-to-be-done
How to know, if success/fail?

4. Challenge
Description of recognized major challenge
How those challenges are solved today

5. Customer process
In what journey is the customer
Where does the customer journey begin from
their perspective)
Where does it end?

6. Buying process and timing
Description of buying process
Seasonal patterns
Budget

7. Should happen (best case)
What are customer expectations
What should happen from customer perspective
(moments of magic)

8. Could happen (worst case)
What could happen (moments of misery)

9. How could we fulfill Customer needs (#3 - #8)
List of ideas

10. Quantified customer needs and metrics
SMARTO -metrics

11. Solution: Partners and technology
What could be done with partners and new technology

12. What business are we in?
Statement in one sentence

Dr Janne Ohtonen 2017 ©

As mentioned earlier, detailed explanation on the above tool is available on **The 5-Star Customer Experience** book. By this time you should have most important information available on what you need to do. The next step is to plan a roadmap on how to get there. And it won't come as a surprise when I tell you that all vendors have their versions of that, too.

If you choose to follow a vendors approach, then it makes sense to use their recommendations on how to build that solution in place. However, you should never give the power of determining your Customer Experience Blueprint to vendors alone as they come and go. Every business should hold themselves only as accountable for CEM strategy and outcomes.

The roadmap for implementing a Customer Experience Management framework will hugely depend on the context, resources and capabilities of your business. The following 5-step approach contains a clear roadmap that works for most telecom organisations:

1. Start
- Define mission and scope for your CEM aspirations
- Create a governance model and educate leadership and management
- Start measuring customer, employee and partner perceptions
- Gain internal buy-in and get started

2. Fix
- Get quick wins and reward for success and positive change
- Fix most important pain points to gain credibility for your CEM efforts
- Link customer, employee and process metrics to business KPIs
- Set baseline and minimum desired level for experiences

3. Improve
- Improve end-to-end customer and employee journeys
- Optimise processes and technology using LeanCEM
- Link metrics to business results and successful customer outcomes
- Establish customer-centricity as a permanent part of the corporate culture

4. Differentiate
- Become famous on few things and tell the world about your aspirations
- Know your customers like your friends
- Use predictive analytics to proactively serve customers
- Align employee and customer experiences

5. WOW
- Become a true customer-centric business
- Grow by customer advocacy and sustainable revenue
- Excel at both customer and employee experience and engagement
- Live every moment by the company's customer-centric mission and values

The level of difficulty rises as the roadmap develops. It is hard to say in this book how long it would take your business to get from step 1 to 5, as it depends on your circumstances and appetite. But in typical telecommunication business, it would be safe to say that it will take at least one year per step. If you already have CEM activities in place, you may be quicker then. Or if you have oceans of change resistance and a bigger transformation is needed, it may take longer. Either way, create your SMARTO[77] roadmap and get on with it!

---

[77] SMARTO = Specific, Measurable, Achievable, Realistic, Time-bound and Outside-In

EPIC CUSTOMER JOURNEYS
- THE LORD OF THE RINGS STYLE

"Getting service right is more than just a nice to do; it's a must do. American consumers are willing to spend more with companies that provide outstanding service — ultimately, great service can drive sales and customer loyalty."
- Jim Bush, A Famous Coach

Managing customer journeys is an integral part of excellent Customer Experience Management. Companies that don't take customer journeys seriously will not be performing well in CEM, because it is the best way to understand how the customers go from their desires to the fulfilment of those desires, through all the experiences, products, services and interactions that are needed along the way. Customer experiences are fundamentally shaped by customer journeys, which on the other hand can be influenced by CEM.

Service design has become a favorite concept in the telecommunications sector. There are also a great many discussions online about the differences between service design and customer experiences. In this book, as well as in *The 5-Star Customer Experience* book, these two terms are used to describe the same concept. Some people limit Customer Journey Mapping to niche journeys and, in the same way, others use Service Design with insufficient scope. Thus, we can agree that both service design and customer journey mapping refers to the corresponding CEM tool as the scope of the approach in here covers end-to-end customer experiences.

This chapter is almost like an executive summary of the whole customer journey mapping method described in 'The 5-Star Customer Experience'. If you want the details of this tool, then you should get your hands on that book. For our purpose here, I have created a Customer Journey Canvas, which gives you an easy and practical tool to use with your team. But before we get into that, let's remind ourselves of why customer journey optimisation is paramount for CEM success:

- **Customers who have high effort service experiences are 96% more disloyal[78]**
- **By 2020, customer experience will overtake price and product as the key brand differentiator, says Walker Info**
- **A 1% improvement in first call response equals to $276,000 in annual operational savings for the average call centre[79]**
- **Maximizing satisfaction with customer journeys can potentially not only to increase customer satisfaction by 20% but also to increase revenue by up to 15% while lowering the cost of serving customers by as much as 20%[80]**

To be honest, I could keep running the above benefits list for several pages. But I believe that by this time, it won't be necessary anymore[81]. We can focus more on how to create customer journeys that lead to experiences that make as long lasting impressions as The Lord of the Rings movie. To do that there are several things that you need to make sure are in place when optimising customer journeys[82]:

- **Make everything easy for the customer**
- **Be proactive in solving challenges and adding value**
- **Start with the customer experience and work your way back to the technology**
- **Get rid of customer lockdown**
- **Continuously improve and optimise customer journeys**

---

[78] Source: Corporate Executive Board report

[79] Source: http://www.icmi.com/Resources/Call-Center-Buyers-Guide/S/SQM%20Group

[80] Source:
http://www.mckinsey.com/insights/consumer_and_retail/the_three_cs_of_customer_satisfaction_consistency_consistency_consistency

[81] But in case you want more CEM statistics, visit here: https://www.garyefox.com/ultimate-list-customer-experience-statistics-2017/

[82] Adapted from 'Next-Level Customer Experience In Telecommunications - Avoiding Hassles Is Not Enough' by Oliver Wyman

Let's go through each of the matters above, playing around with the Customer Journey canvas and any ideas it might give. Let's start with making everything easy for the customer. Everyone is busy nowadays, and no one wants to spend time on chores that do not add value. Typically, dealing with one's operator has been such a tedious chore. Digital service providers who are getting the basics right in reducing customer effort and improving usability and experience have significantly reduced the number of price plans and options as well as removing one-off fees and unsustainable revenue sources (e.g. customer penalties).

They have also removed twelve-month terms, bearing in mind that unsatisfied clients locked into long contracts can become serious detractors on social media. Many successful communication service providers have simplified their terms and conditions, writing them in layman's language. Also, today, too many customers are still confused by billing, random credits, unrewarding loyalty schemes, and lack of clarity on what they are paying for.

Thus, another way to reduce customer effort in telecoms is to create a simple invoice showing a single total (including all taxes, fees, charges and discounts) across all the services utilised from that operator. Though previous use case would only help an operator to achieve an entry-level grade customer experience, there are WOW opportunities to be had. Many of today's clients can still be surprised with a zero set-up experience. And even tech-savvy Generation-Y can be positively surprised by an operator's mobile app guiding them through a video-based troubleshooting process powered by artificial intelligence. Using the Customer Journey canvas, you can find the most relevant opportunities to fix the basic experiences and create more WOW experiences.

The second matter you need to consider when working on customer journeys is to be proactive in solving challenges and adding value. Being proactive benefits both sides. For example, telling clients about planned maintenance means reduced effort for businesses (fewer calls to agents) and less dissatisfaction for clients (from spending time in call queues and trying to troubleshoot the issues themselves).

One Latin American operator is known to take a pre-emptive measure to remotely re-start its broadband routers during the night to avoid racking up errors, degrading TV and calls quality. Progressing to a differentiated customer experience requires service providers to go the extra mile and do more than just giving reactive information. They should solve problems in such ways that are relevant and proactive.

As an example, customers of customer-centric digital service provider might receive a notification like this: *"Hi Peter, we noticed your broadband connection is down at the moment. Our engineers are already on it, and while we fix it, please use mobile phone tethering, which you can find in our mobile app. As you frequently use Netflix, we activated a zero charging for Netflix on your mobile so you can watch your favourite series for the next 30 days on us. Sorry for the inconvenience, we will sort it out soon."*

Those multi-play operators who value their customers are already taking action to implement such pro-active measures. As an example, a Western European service provider automatically monitors broadband throughput and opens a service ticket internally as soon as it drops below 80% of what has been promised to the customer.

Let's continue playing with the Customer Journey canvas. Another idea for a revolutionary approach would be to offer an automatic migration to the latest pricing, eliminating the customer hassle of switching plans while also reducing service provider's effort of maintaining legacy products.

"You've got to start with the customer experience and work back to the technology – not the other way around"?

- Steve Jobs, Apple

Unfortunately, too many companies in the telecom sector have missed the above quote as they are still focusing on technology offers, such as *"6GB of mobile data over a 4G/LTE connection at up to 40Mbit/s for $20 a month"*. This kind approach is very inside-out focused and problematic for many reasons. Firstly, it does not speak to what clients consciously want or unconsciously need, such as *"be online on Facebook with friends and WhatsApp 24/7 to my boyfriend"* or *"watch Netflix uninterrupted when I am commuting to work"*.

Technical specifications may seem important to communication service providers, but do they satisfy the wants and needs of customers? Would you know how many episodes of Breaking Bad you can watch with that 6GB? Such an approach also limits the use of service providers' critical assets, such as their fixed and mobile access networks, to improve customer experience. Digital service providers typically keep an eye on data volumes as for their primary differentiator regardless of when the data is used.

Some service providers restrict data volumes or block specific applications at times of heavy network load. However, a better use would be made of networks by tailoring their offer from a customer perspective. As an example, a proposal to *"watch as many videos as you like and we will set the optimal resolution for you"* would sound better for the customer and maximise the use of the network for an operator.

Building on that, this kind of approach could provide opportunities to offer a great customer experience at lower operational cost when there is spare network capacity available: *"Make free mobile HD video calls for the next two hours"*. These are the kinds of ideas you can get by playing around with the Customer Journey canvas.

Communication service providers are notorious on locking customers down on long contracts. It is understandable when customer acquisition is so expensive, and dog-eat-dog competition has driven value out of the industry. Still, no one likes to be locked down to a service provider. In fact, clients increasingly expect to be able to adjust the services they consume in a flexible and personalised way. Telecommunication customers do not live in a vacuum.

New cloud and Software-as-a-Service (SaaS) players such as Amazon Web Services, Google Cloud, and Rackspace have changed their expectations as they allow clients to change service features in real-time. And that is without the burden of fixed-term or punitive contracts, out-of-bundle costs, or unused quotas. Some firms offer telecom products with greater flexibility than the usual tiered bundles. As an example, Virgin in the U.S.A. (in collaboration with Walmart chain) launched the 'Data Done Right' plan, enabling clients to share data volume among multiple lines.

The Australian operator, Yatango Mobile, offers a slider adjustor, giving clients the flexibility to choose voice and data volumes as they like. Yatango Mobile also proactively recommends beneficial changes that customers can make to their existing plan, based on real-time usage, with no fixed term. Allowing customers to configure their product or service themselves means no more legacy tariffs or marked-up out-of-bundle charges to punish clients.

By enabling the adjustment of unit prices per service for all customers, service providers could minimise reconnections costs of existing clients, saving on Subscriber Acquisition Cost (SAC). However, in this kind of approach, to limit dilution of the Average Revenue Per Unit (ARPU), service providers would need an upsell strategy to compensate the change. They could, for instance, adjust all sliders for existing customers to keep them on the same ARPU while letting them choose "pay less" or "get more".

As a final point, customer journeys need to be continuously improved and optimised. What is noticeable is that best-in-class telecommunication companies optimise end-to-end customer journeys, not just few of the steps. They also have visibility into how those individual steps along the journey impact the end-to-end journey. Here's an example of how to do that using the Voice of Customer data available from each of the steps[83]:

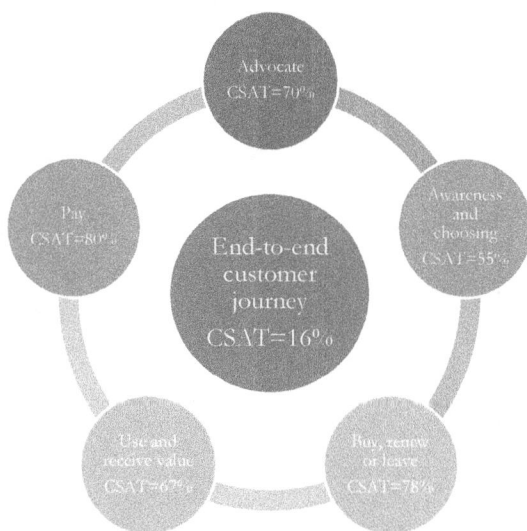

As you can see from the above example, the end-to-end satisfaction declines rapidly along the journey as each of the steps erodes the customer's satisfaction. And the best medicine for that is to start working on the customer journeys to improve and optimise them. Here's a practical tool for you to get started and you can get more details on how to run workshops to use the tool are available on 'The 5-Star Customer Experience' book:

---

[83] These figures are for illustrative purposes only. Calculation: 0.7 * 0.55 * 0.78 * 0.67 * 0.8 = 0.16

DR JANNE OHTONEN

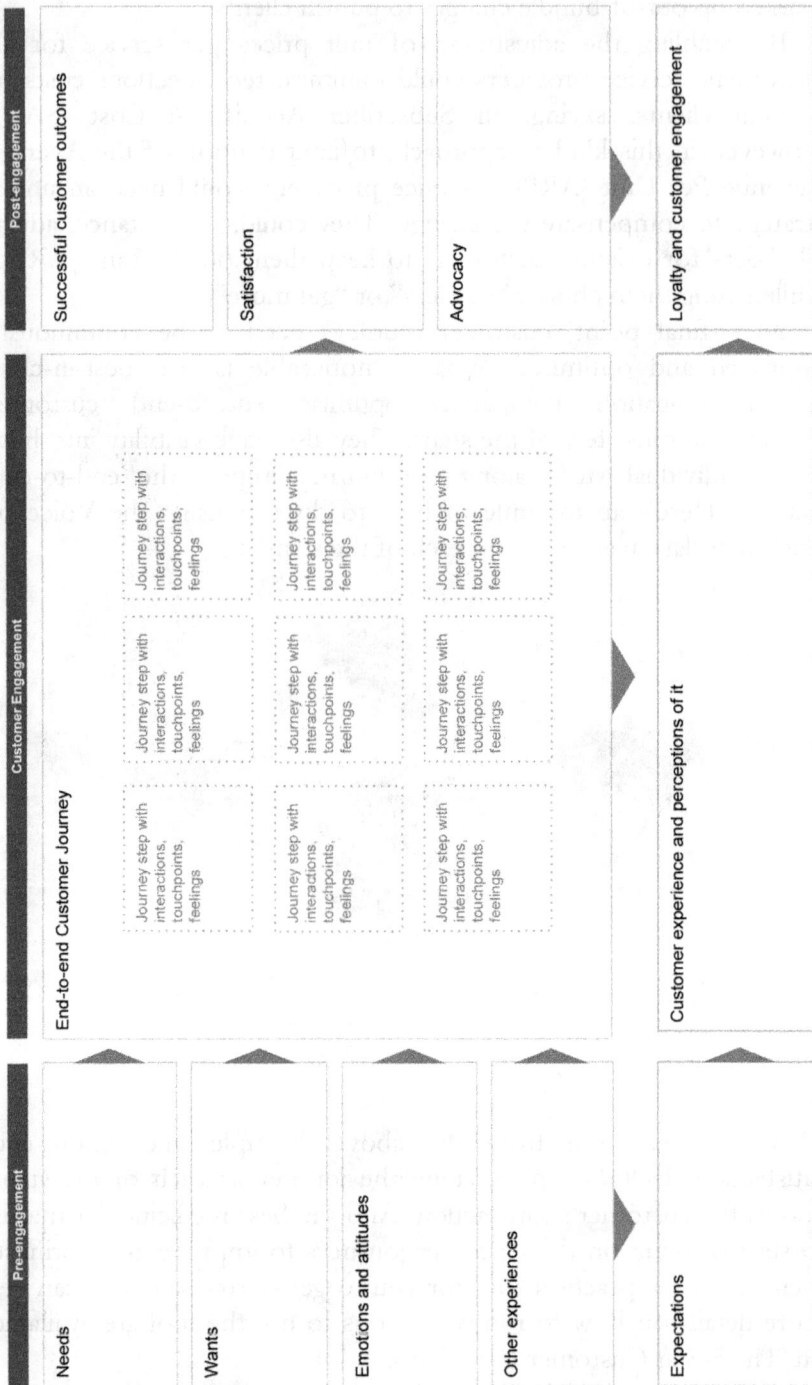

The final thing to do is to make sure that these important plans won't be forgotten. One of the easiest ways to do that is to assign each customer journey to accountable members of staff in the organisation. Here's a simple table that you can use as an example to design how it could work in your company:

JOURNEY	IMPROVEMENTS	METRICS	PERSON
Awareness and choosing	Inbound marketing, giving value from 1st interaction	Brand equity, Leads	VP, Marketing
Buy, renew and leave	Simple offers and sales journey	Sales conversion	VP, Sales
Use and receive value	Product reliability improvement	QoE, Number of service calls	VP, Product
Pay	Simplify and clarify invoicing	Number of calls about invoices	VP, Finance

This chapter has looked into one of the most important elements of a successful Customer Experience Management framework for a telecommunications company. It has been merely a scratch on the surface, but hopefully a useful one. As mentioned already, you can find more information on both Customer Experience Blueprint and Customer Journey Mapping from 'The 5-Star Customer Experience' book. Don't let the simplicity of the tools presented here fool you as they have tremendous value to provide for your business.

DON'T BE AN EMOTIONAL TRAIN WRECK

"It's easier to love a brand when
the brand loves you back."
- Seth Godin, Author, The Icarus Deception

This book is a weird one. Though it is about Customer Experience Management in Telecommunications, it has this theme of marrying customers as its overarching analogy. And since this chapter focuses primarily on emotions and feelings (you know, that fluffy stuff), I suppose it is an appropriate way to get started with a talk about emotions in marriages? Speaking of which, are you aware of "The Nothing Box"[84] every man has? If you are a man, I bet you are. And if you are a woman, I bet you have come across it, but may not have realised that it is a real thing! So, what is The Nothing Box and what does it have to do with CEM?

Mark Gungor (a person vividly expressing what The Nothing Box is on YouTube as referenced in an earlier footnote) thinks marriages are an energy-giving institution. But for some, the marriage has become an energy-sucking one. And that's why many of those who feel the energy drained from their lives due to marriages try to instil in those wanting to get married a huge burden of having everything ready before getting married. And that advice doesn't always work, ending too many marriages prematurely.

Yet, it does not have to be that way, if you do it right! Very similarly, the purpose of customer experiences is to produce happiness and value for the customers and for the business. It should not be a tedious task that is done with a huge amount of advice and maturity levels that can never be reached. Same way as in marriages, it does not have to be that way. *"If you do it right, marriage [or customer*

---

[84] Mark Gungor – The Nothing Box video on Youtube https://www.youtube.com/watch?v=SWiBRL-bxiA

experience] can be the closest thing to a Heaven on Earth, and if you do it wrong… well, you fill in the blanks", Mark says.

Mark speaks about how women's and men's brains are very different emotionally. I would like to extend this same concept to customers' and service providers' mind-sets. Mark explains how men have a box for everything: car, money, job, spouse, kids, and they even have a little box for the mother-in-law somewhere in the basement. And the rule is that these boxes do not touch each other!

Men like to deal with one box at a time. Now women come from a completely different place with their brains. They have a big ball of wire that connects to everything. The money is connected to the car, the car is connected to the job, and the job is connected to the money. It is almost like an Internet of Everything! And their brains are powered by the energy we call emotion.

Now think about this in the context of customer experiences. Customers are like women (though they can be men too!) when it comes to experience. No experience is gone through in a vacuum, and everything is connected to previous experiences, expectations, current emotional state and many other things. What about the service providers then? They are much more like men. They have a small box for everything, and those boxes do not touch.

There's a box for selling stuff, using products, engaging with services, getting support when something is broken, paying and pushing self-promoting materials. Are you starting to see how the analogy of the male and female brains is very relevant to differences between customer and operator mind-sets? If you are, we can go back to The Nothing Box as it will make much more sense now.

Men have a very special box, called The Nothing Box. It is their favourite box. If a man has a chance, he will go to his Nothing Box any time. That's why he can do brain-dead stuff like fishing, channel surfing or just sitting for hours on. And of course this drives their wives nuts as she will come up and just shout *"stop, you can't possibly be watching anything"*, while the man goes *"uh?"* with this uncalled interruption to this Nothing Box time.

Mark claims that the University of Pennsylvania made a study that proved the men can think about absolutely nothing and still breathe. Women can't do it as their minds never stop. The Nothing Box is an enigma for them, and it drives them crazy as nothing is more irritating for them than to witness a man doing nothing! So, let's start pulling this analogy back to customer experiences. While service providers are in this vegetative mindset, focusing on nothing, the customers are going through a vast range of emotions.

Let's take a couple of examples. How many times have you as a customer been very frustrated with a wait time? Maybe you waited to be served in a branch or just had the famous Kevin Bacon Buffer Face while your mobile was downloading something from Netflix? Now, what was your service provider doing meanwhile? Enjoying their time in The Nothing Box.

If you work for a service provider, how many times have you actually felt concerned about a customer waiting for YOU (or your business, if you don't do anything directly yourself to customers)? I bet not often. And that is just one example of emotional issues such mindset can cause. The hope is not all lost though, neither in marriages nor customer experiences. As we are now aware how emotions don't only affect how we feel, but also how we behave, we can do something about it.

"Negative emotions don't even have to be extreme, the absence of positive ones can be enough to cause customers to vote with their feet and go elsewhere. Companies that aim for emotional connection beat their competitors by 26% in gross margin and 85% in sales growth, with customers feeling more engaged and appreciated."
- Forrester Research, 2017

There are very practical ways of understanding customer emotions and doing your best to influence them positively. You can refer to 'The 5-Star customer Experience' book for detailed instructions on how to do that as part of customer journey mapping. Additionally, the Customer Experience Blueprint includes guidelines for how to understand customer emotions on a strategic level. A third great source of insights into customer emotions is through a Voice of Customer programme that utilises both structured and unstructured data. Since the two first approaches are already discussed in details in the other book, let's look into using VOC with unstructured data to understand emotions with telecom customers.

With one service provider, we used survey verbatim, social media chatter, call centre transcriptions and direct customer feedback to understand what kind of emotions customers had in different key touch points. One big upside of unstructured feedback is that it allows customers to tell their thoughts in their own words. Though not all customers will express clearly detectable emotions in their feedback, we have still found that in most sources of unstructured feedback a large number of clients will give indicators into their emotional states along the customer journey. By using advanced text analytics and natural language processing algorithms trained to detected relevant customer experience emotions, you can identify when customer emotions are present in the customer's comments at a thought phrase level. Here are results from such analysis from the previously mentioned service provider:

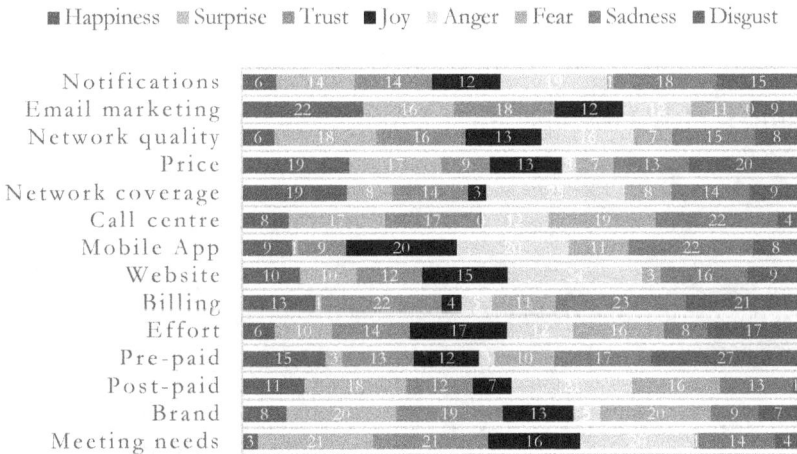

■ Happiness ■ Surprise ■ Trust ■ Joy ■ Anger ■ Fear ■ Sadness ■ Disgust

Touch point	Happiness	Surprise	Trust	Joy	Anger	Fear	Sadness	Disgust
Notifications	6	14	14	12		18	15	
Email marketing	22	16	18	12		11	9	
Network quality	6	18	16	13		7	15	8
Price	19		9	13	7	13	20	
Network coverage	19	8	14	3		8	14	9
Call centre	8	17	17		19	22	4	
Mobile App	9	9	20		11	22	8	
Website	10	12	15		3	16	9	
Billing	13	22	4		23	21		
Effort	6	10	14	17		16	8	17
Pre-paid	15	3	13	12	10	17	27	
Post-paid	11	18	12	7		16	13	1
Brand	8	20	19	13		20	9	7
Meeting needs	3	21	21	16		14	4	

The above chart gives insights on dominating emotions in different touch points across the customer journey. What happens here is that the operators are not left with The Nothing Box anymore. It gives the "woman's view" into the energy of emotions. You can use this information to prioritise your efforts on which areas to improve. It will help to optimise the desired emotional impact based on the Customer Experience Blueprint, Customer Journey

Maps, and Voice of Customer insights.

"People will forget what you said,
people will forget what you did,
but people will never forget
how you made them feel."
- Maya Angelou, an American Poet

We have already talked about customer expectations and customer journeys. That was more from strategic and functional perspectives. You are now advanced enough to add in the emotions and consider them as part of your CEM Blueprint as well as journeys. The way you can maximise the potential for desired feelings (remember that, in the same way as actually managing customer experiences, it is tough to control someone's feelings, though not impossible), is by being aware of feelings at all stages and factor them in your user experiences and journeys. Here's a simple framework for setting emotional customer expectations:

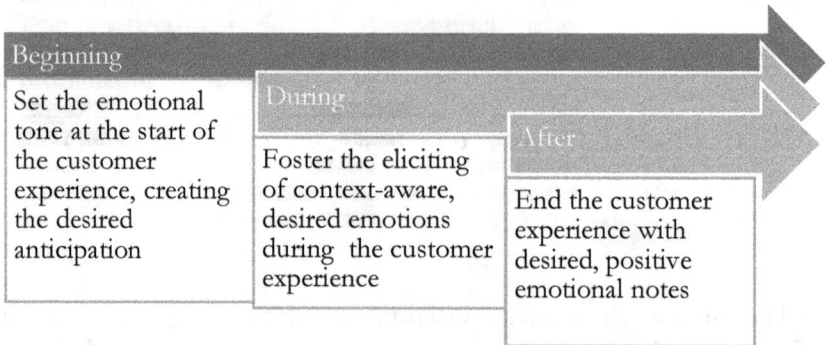

Beginning	During	After
Set the emotional tone at the start of the customer experience, creating the desired anticipation	Foster the eliciting of context-aware, desired emotions during the customer experience	End the customer experience with desired, positive emotional notes

Note that the emotions that have been elicited in the previous stages of the journey will linger through the rest of the journey. That is why I say this quite often when talking about the disappointments life may bring:

"Disappointment is the difference between
expectations and the reality."
- Dr Janne Ohtonen, Customer Experience Leader

Finally, let's discuss which emotions are important and how you can take practical steps to enforce those emotions with your customers. Bob Thompson looked at data and found out that certain emotions are more important for customers than others[85]. As a headline finding of its own, it's not huge insight for you. I have conducted emotional studies in various large companies and conclude that the most important emotions vary for every business as they have different kinds of customer bases. So, to truly know what emotions matter for your customers, you will have to conduct a study of your customer base. Meanwhile, we can look at emotions from the average perspective and highlight emotions that are most likely to be important:

- **Satisfaction is three times more likely to be the most important emotion than the second one**
- **Feeling safe and being able to trust the service provider are likely to be in the top 3 most important positive emotions**
- **Frustration and disappointment are most likely the strongest negative feelings towards service providers**
- **Feeling disrespected and angry are likely to be in the top 3 most important negative emotions**

As we know, telecom clients who feel that they have been treated unequally are likely to leave. A great example of such a situation is how new customer are treated compared to existing customers (e.g. giving half-price deals for new customers to join while increasing prices for existing customers). Considering such basic emotional

---

[85] http://customerthink.com/satisfaction-is-dead-not-its-the-most-common-emotion-in-great-customer-experiences-research/

factors helps operators to reduce customer churn and dissatisfaction. Simple things like making the same offers for existing and new customers, eliminating small print, and keeping brand promises should be self-evident. Let's take an example of enforcing feelings of being safe and able to trust the service provider during the sales and renewal processes:

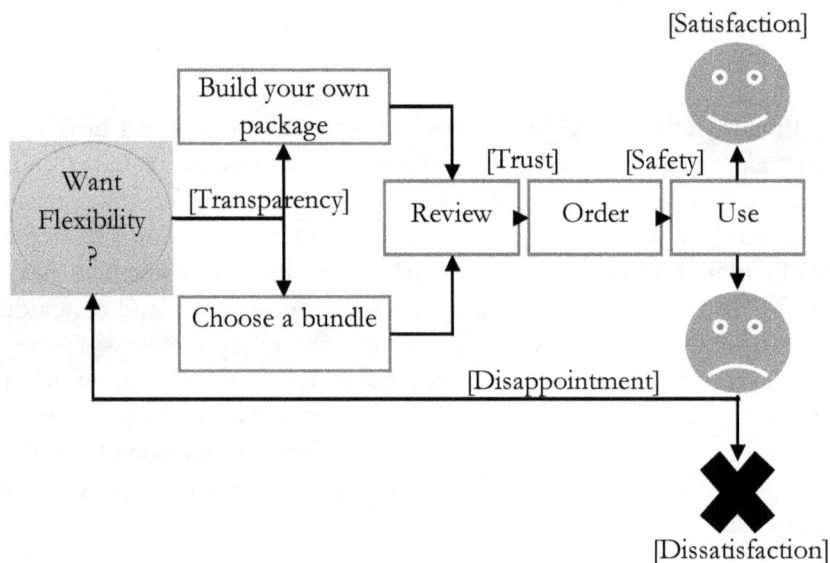

Any telecommunication service provider who is serious about delivering great customer experiences needs to undertake emotional mappings of their customers. They are fully transparent to the customer, regarding what is going on. A European communication service provider, for instance, developed a mobile app to display the status of a customer's enquiry in real-time, greatly reducing inbound calls for status checks. Also, companies like Google and Skype show the real-time technical status of services on their websites.

Trust is one of the most important emotions for telecom companies to create and it is good to start with basics: it is important for clients to see their service status, the service level they can expect, and features or services they might have to pay for. Netflix has been successful doing that by first establishing a customer-centric culture and second by empowering its agents to interact with clients in a very appreciative and personal way. This has led to a WOW customer experience that has helped to deepen the customer relationship straight from the start.

Communication service providers will have to connect with and learn about their customers not only from functional bit also from emotional point-of-view. For instance, by connecting with its customers on Facebook, a service provider could enrich its Customer Relationship Management data and enable its agents to relate to a customer's personal life (where appropriate) more accurately (with knowledge about hobbies and interests, reference to recent holiday photos, and so on). To offer a personalised experience, service providers need to be able to participate at any level on the personal appreciation scale – from 0% in anonymous web communities run by customers to 100% in personal service models for high-value customers and do that at massive scale. How does that make you feel?

VOICE OF CUSTOMER – CAN YOU HEAR IT?

"If you make customers unhappy in the physical world, they might each tell 6 friends. If you make customers unhappy on the Internet, they can each tell 6,000 friends."
- Jeff Bezos, Founder, Amazon

Social media has taken hold of our lives for over a decade already, and it is just getting bigger as the younger generations will have been born with it. Every second, on average, around 6,000 tweets are posted on Twitter, which corresponds to over 350,000 tweets sent per minute, 500 million tweets per day and around 200 billion tweets per year[86]. Facebook has more than 2 billion active users globally every month[87].

That is shedload of sharing, and part of that sharing is about our businesses, too, as Jeff Bezos describes in the above quote. And this is what a robust Voice of Customer (VOC) programme looks to understand and harness. There are still too many telecommunication companies who think that VOC is just about sending out long surveys to their customers and about responding to queries. But it is so much more than that:

- **If a telecom company responds to customer's concern within an hour, 75% of those customers are likely to buy again, but if they respond later than that, only 20% will consider buying from them again (Forum Corporation)**
- **Highly engaged customers bring a 23% increase in share of wallet, profitability and revenue as compared to the average**

---

[86] http://www.internetlivestats.com/twitter-statistics/

[87] https://www.statista.com/statistics/264810/number-of-monthly-active-facebook-users-worldwide/

customer (**Harvard Business Review**)

- Engaged customers repurchase 57% more frequently than those less connected, and they are 42% more likely to be a customer a year later (Carlson)
- Businesses with higher levels of engaged customers outperform peers by 26% in gross margins and 85% in sales growth (Bain and Company)
- 84% of clients who leave do so due to poor service and of these only 4% proactively complain (Forum Group)

This book is heavy on statistics, but I only share those with you that will help you to convince the rest of the business on the importance of CEM. The above statistics show why having a comprehensive Voice of Customer programme that addresses customers concerns across all channels and data is so important. And it is an amazing opportunity to increase customer engagement.

You might be asking, what is Voice of Customer anyway? It is a systematic management programme (part of CEM) designed for acquiring business insight about customers and what is important to them. It works on knowing and understanding the customers better on an individual level, focusing on customer's needs, wants, expectations - and most importantly on perceptions of your brand, products, services and the market.

I am sure you have become aware that most things in CEM, including VOC, have extensive definitions and focus areas. And that is why people working on customer experiences need to educate themselves and widen their perspectives continuously. It is human work at its best. That is also why benefiting from Voice of Customer requires more than the occasional sending of long surveys. It requires a wide-ranging, strategic and ongoing dedication to hearing, listening, understanding and acting upon the VOC through a formal programme built upon the following principles:

- Producing successful customer and business outcomes – how will you make sure that investment in VOC programme will yield the results and doesn't just become an academic exercise?
- Proactive listening – how will you build an always-on system providing all customers with the opportunity to share

compliments, complaints and comments about their experiences with your products and services?

- **Background monitoring** – how will you set recurring, unnoticeable and systematic ways of tracking changes in customer and business outcomes, their leading indicators, and the important drivers?
- **Tuning in through the organisation** – how will you ensure that everyone in the business from the board to janitor is listening to the same customer channel?

The above list of VOC principles is reasonably simple, but hard to implement. Based on having done this in dozens of organisations, it seems to take anything between one to three years, depending on the appetite of the organisation. But by no means is it an endeavour to be taken lightly, though it is more than worth it as discussed earlier.

This book has shown you several frameworks (essential supporting structures) for CEM and its subcomponents. Voice of Customer needs its framework also as it is a complicated matter. It is not sensible to go into the details of designing and implementing a VOC programme in this book as there are already several books written about it already[88]. However, I will share here information that I have found to be missing from those books and is essential for a successful VOC programme in the telecommunication sector.

Let's start with how we will get customer feedback and use it to ours and customer's advantage. The main avenues for telecom companies to explore are:

---

[88] Just search for "Voice of customer" on Amazon or any other reputable bookshop

We ask our customers
- Relationship surveys (customers, employees, partners)
- Transactional surveys (touch points, win/loss, churn)
- Ad-hoc surveys (brand, marketing, product, service)
- Focus groups, communities, discussion forums

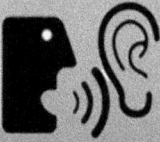

We listen to our customers
- Complaints
- Feedback forms
- Employee and partner feedback on experiences
- Unstructured data and feedback (emails, calls, forms)

We monitor customer experiences
- Gap and driver analyses
- Social media, reviews, blogs, analysts, big data, competition, OTT players
- Behaviours, lifetime value, ARPU, attrition
- Quality of Experience monitoring

We learn!
- Analytics, prediction and benchmarking
- Data mining, machine learning
- Text and semantic analytics
- Enriching data with integration to other sources, e.g. CRM, ERP, eCommerce, OSS/BSS

We take action!
- Case management for unsatisfied customers
- Company- and market-level action
- Prioritised roadmaps for improvement plans for all business areas and roles

How well are all five areas covered by your company today? I have a guesstimate that some of the topics may be included properly, but especially monitoring customer experiences and taking actions based on insights may be struggling (at least that is a typical situation with other telecom players in the market). The impact your VOC programme has depends on what maturity level it is on and how well it is able to perform the above five steps consistently. You can use the maturity model below to evaluate where you are today and what you could gain more by increasing the organisation's performance on the Voice of Customer:

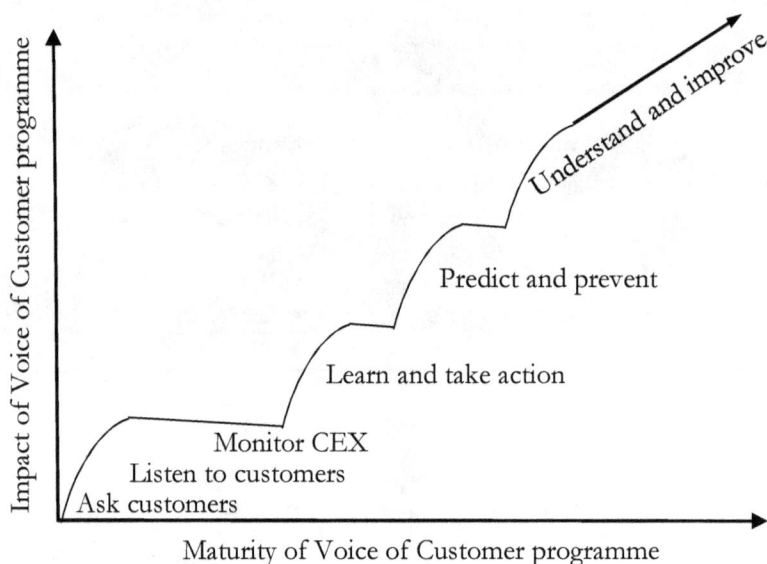

As the above VOC maturity model illustrates, your current activities will eventually lead to a decline if the next level of operation is not started at the right time. Companies who are in such a state do not receive enough benefits from a VOC programme and are at risk of stopping doing them. I have personally rescued several companies where the senior leadership told me how useless customer feedback is.

I felt sorry for their customers and therefore I saw it as my duty to help those companies to revive their VOC efforts and to set them on a road that would take to the "Holy Grail" of VOC, that is, continued increase in customer understanding and meaningful action to improve. Also, reflecting back to what I said about maturity models, the above is not intended to be one, but more to show on a conceptual level where any communication service provider needs to get to if they want to stop losing customers and money with suboptimal Voice of Customer programmes.

In my previous book, *The 5-Star Customer Experience*, I had a section that shed light on how different departments in a company could benefit from the Customer Experience Blueprint and Customer Journey Mapping. That part of the book has inspired numerous discussions in corporations, and therefore I thought to include similar kinds of chapter for the Voice of Customer, too. Many leaders ask, *"What's in it for me?"* and the following list will help you to answer that question.

Let's start with our beloved **senior leaders**. For them, a VOC programme becomes a vital tool for shaping the company's strategic vision as well as the Customer Experience Blueprint. Profoundly understanding clients' needs allows service providers to form a roadmap to position the company to fulfil those requirements better.

For **marketing** to have maximum effectiveness, it must be relevant, contextual, and attractive to potential customers. A VOC programme facilitates the development of effective market positioning and messaging by incorporating the language and the viewpoints of prospects. Marketing will also benefit from the identification of buying decision factors and customers' ways to express their needs and challenges.

Understanding the leading indicators and drivers of customers' perspectives toward a service provider allows **sales** management to proactively detect, and take pre-emptive action against, negative trends in business outcomes. Unlike typical sales management systems, a VOC programme will enable sales professionals to see beyond *"what is going on"* with insights that show *"what might happen"* and even more importantly, to understand the reasons why.

For most telecommunication service providers, the primary interaction point between the company and the customer occurs when service is needed. It may be a question to be answered, a problem to be solved, or confusion to be clarified. And as many operators are still in the Stone Age when it comes to digital services, it also might be to make an order, an inquiry about an order or to renew a contract. Getting this interaction right is a valuable opportunity to strengthen the relationship. A well-designed Voice of Customer programme allows a service provider to understand what works and to monitor customers' perceived satisfaction with those interactions.

In a rapidly evolving marketplace, firms must continually improve existing **products and services** as well as identify future solutions to customer's challenges. A VOC programme can contribute to the achievement of both aspirations. And the best part of it all is that, in the process of doing that, VOC allows an operator to broaden the depth and breadth of their relationship with the customer, resulting in improvements in share-of-wallet.

As Peter Drucker said already in 1954, the reason for a business to exist is to gain a customer. A VOC programme can help **finance** professionals to see that all revenue comes only from two sources: what customers spend with the enterprise today, and what they are likely to spend in the future. These sources of customer equity are known to be profoundly influenced by customers' perception of - and experience with - the company. Thus, identifying and monitoring the quality of such opinions and experiences is as paramount to the sustained health of the business as identifying and tracking cash flows.

Maybe you wouldn't think that customer feedback can help **human resources** to perform their job better as well, but it is true. A Voice of Customer programme provides insights into those specific employee traits that contribute to the development or the destruction of customer relationships. This knowledge can be used in hiring decisions, in performance management, and in designing training programs.

Now that we have established the need for Voice of Customer in telecom companies throughout various departments, what would be your guess on how many of them actually use that insight? Not many it seems[89]:

---

[89] Source: The ROI on Customer Feedback by Zabin J. from the Aberdeen Group

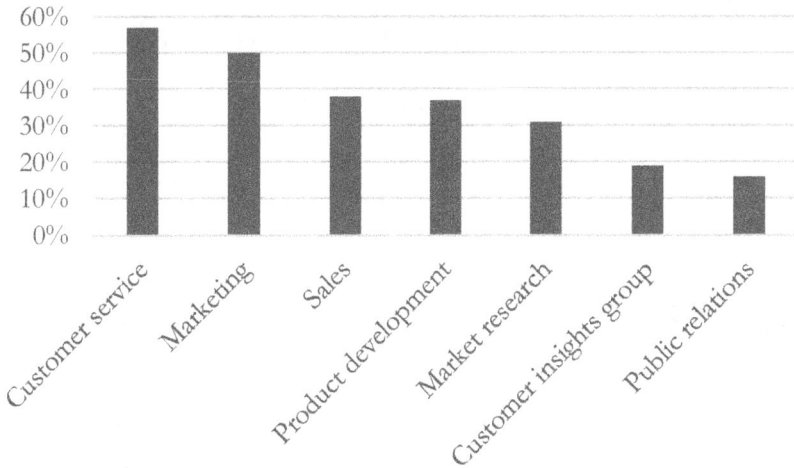

The above chart is only useful for you from the perspective of challenging what these percentages might be in your business? I am willing to bet that a significant amount of accumulated profits (not from this book though as they go to charity) that you will find departments who are barely utilising the insights generated from your Voice of Customer efforts. When you ask them, remember to be clear about what you ask. Whether they read the reports is very different to taking action based on them and then following up to see how it worked.

"We have only two sources of competitive advantage: the ability to learn more about our customers faster than the competition, and the ability to turn that learning into action faster than the competition."
- Jack Welch, former CEO of GE

The scope of this book is not to teach you how exactly to design a Voice of Customer programme. For that, you need to refer to other books or get a hold my team or me (we'll be happy to help you out).

To move from theory to practice, I would like to share with you the most used customer metrics for telecommunication companies and why they are so popular. Then I will share with you my template for measuring VOC through a relationship survey, and I will finally end with the practical process on how to deal with the insights you get from it. Let's start with the most common customer metrics that telecommunication companies use:

METRIC	DESCRIPTION	QUESTION	WHY
Net Promoter Score, NPS	Measures how likely customers are to advocate the business, brand, product, etc. to their peers.	How likely are you to recommend [...] to a friend, relative or colleague?	Customers will only recommend something to their peers if their emotions and logic say it makes sense to do so.
Customer satisfaction, CSAT	Measures how satisfied customers are with the aspect.	Overall, how satisfied are you with [...]?	If customers are not satisfied, they will take their business elsewhere.
Customer effort, CSAT	Measures how easy it was for customers to deal with the organisation.	Overall, how easy is [...] to deal with?	If things get too complicated, people will get frustrated.
First call resolution, FCR	Measures how often customer issues are resolved with one interaction.	Has your issue been fully resolved in the first contact?	Issues should not happen in the first place, so you should deal with them promptly when they do.

The above metrics are only examples of typical Voice of Customer metrics. They don't consider unstructured data, other data sources, non-survey feedback, etc. But it is good for everyone in your organisation to be aware of the above basic metrics.

Let's put the above metrics to practical use. I promised you a template for a relationship survey, so here it comes. Whatever you paid for this book has just been returned to you thousand fold as many vendors sell such information for big bucks. Naturally again, one size does not fit all, and you may need to reformulate the template below to suit your target audience and business:

- How likely are you to recommend [BRAND / PRODUCT / SERVICE] to a friend, relative or colleague?
- Overall, how satisfied are you with [BRAND / PRODUCT / SERVICE / CHANNEL / TOUCHPOINT]?
- Overall, [BRAND] makes it easy for me to handle my issues with them
- How well do [BRAND] products meet your needs?
- How enjoyable is [BRAND] to do business with?
- To what degree do you trust that [BRAND] will take care of your needs?
- How likely are you to consider purchasing [PRODUCT] from [BRAND] again?
- How likely are you to change your current [MOBILE / BROADBAND / LANDLINE / TV] service provider within the next 6 months?

Accompanying the above questions with proper question logic, logic that will ask open-ended questions when customers are not happy, will equip your Voice of Customer programme with enough both quantitative and qualitative data so that you can take tangible action to improve the business. VOC surveys don't always have to be long and tedious as the above is already enough to get started for the first year. And that leads us to our final point of Voice of Customer: what should you do with the feedback you get?

A process called close-loop refers to taking the customer feedback from a VOC programme and using it to benefit both customer and the business. As we have already discussed, once we have the customer feedback through and are listening to them, asking and monitoring, we learn from that feedback and take action. Service providers can take action on many levels, some of which are discussed through the use cases in the next chapter.

Meanwhile, you have two main areas to focus on: operational and strategic close-loop. The operational close-loop focuses on the individual customer who the feedback is related to. This might mean calling back to the customer to get more information or to let them know that their feedback has been heard and the issue has been resolved. The strategic close-loop focuses on pulling together root causes for problems that are recurring.

It takes care of not repeating the same problems. If you think about this from a perspective of a harem, operational close-loop will focus on nagging of one of your wives/husbands while the strategic close-loop would fix systematic issues in your harem operation, ensuring most of the wives/husbands you have in there are happy. And on that thought, we can move into our next topic.

QUALITY OF EXPERIENCE

"We focus a lot on the quality of experience, speed, and reliability. It's not sexy from many people's perspective, it's not glitzy in the feature set, but it's what people come to rely on."
- Jan Koum, CEO, WhatsApp Inc.

It is practically impossible to talk about customer experiences in telecommunications without considering the Quality of Experience (QoE) concept. It would not be difficult to argue that QoE is automatically included in CEM as network quality is one of the biggest drivers of customer satisfaction as seen at the beginning of this book. Thus, QoE is a more technical view of how telecom service providers produce value for their customers.

We will go through use cases related to Quality of Experience later in this book. This chapter is not intended to be an in-depth view to QoE as there is already readily available a great deal of information on the technical implementation of it (and my team is more than happy to advise you on this). However, it is worth our time to explore the concept, especially about Customer Experience Management. Let's start with that QOE is. In 2013, Qualinet[90] defined QoE as:

---

[90] For more details, see European Network on Quality of Experience at http://www.qualinet.eu

> *"The degree of delight or annoyance of the user of an application or service. It results from the fulfilment of his or her expectations concerning the utility and/or enjoyment of the application or service in the light of the user's personality and current state."*
> *- Qualinet, 2013*

It is not hard to see the link between the above definition of QoE and CEM. Like CEM, the QoE is an emerging multidisciplinary field based on cognitive science, social psychology, finance, and engineering science, centred on understanding overall individual quality requirements[91]. Quality of Experience is a concept most closely linked to telecommunication and video services, thus having a narrower scope than CEM. QoE is known to have emerged from the idea of Quality of Service (QoS), which focuses on objectively measuring telecommunication service parameters (e.g. average throughput and packet loss rates).

The difference is that QoS measurement is mostly not related to a customer, but to the media a customer consumes or to the network itself. QoE, however, is closer to Voice of Customer as it is a subjective measure from the client's perspective of the overall quality of the service provided. Also, QoE is strongly related to but different from the field of User Experience (UX), which also focuses on users' experiences with services.

In contrast to UX, the purpose of improving QoE for customers was more strongly motivated by economic needs than usability and aesthetics. For our purpose with Customer Experience Management, we can put QoE as part of the VOC technique and include it as a data source for understanding how customers use our services.

Continuing our theme of marrying customers, we can see the VOC as the overall feedback mechanism and pulse check for the status of our marriage. It reveals how we are doing with our spouse and what are the areas to work on. In that QoE is an area that measures our performance in the area of the network, which is the

---

[91] https://en.wikipedia.org/wiki/Quality_of_experience

vehicle we use to deliver the value to customers.

Today, more people than ever are choosing a smartphone or other mobile device as their platform of choice for watching video content from the Internet due to its convenience. Some people use their devices to catch the latest news highlights, while others view entertainment shows on the train during their commutes. Especially in urban areas, many use a Wi-Fi connection in the local coffee shop to connect to Netflix or YouTube on their tablets.

Never mind the type of a device, content viewed, or network used for access, everyone has basic expectations about the viewing experience. The level of expectation varies among customers, depending on factors such as the type of device being used for viewing, the amount of money spent on the service, and kind of content being requested. Thus, it is the cumbersome job of digital service providers to match the level of quality to their customers' expectations of the experience.

This is where QoE specifically comes to play. It looks at an operator's offering from the customer's perspective, and asks, "*What combination of products, services, and support will provide you with the perception that the total solution is providing you with the experience you desired and expected?*" It then continues to ask, "*Is this what the digital service provider has actually provided?*" And if not, "*What kind of changes need to be made at the service provider's and your end to enhance your perception of the Quality of Experience?*" Asking these questions, QoE provides an evaluation of customer's expectations, feelings, perceptions, cognition and satisfaction concerning a particular product, service or application. It can then further be used to optimise the experience and therefore the customer satisfaction and lifetime value.

Quality of Experience is vital for customers. Conviva[92] asked from its consumer base, what they consider as a most crucial factor in choosing a service provider for a video service. The QoE was reported to be most important with a 35% stake, while other factors like price and recognisable brand came behind wit 23% or less. And the need for high quality connections is needed more than ever:

---

[92] Source: Consumer Quality Report by Conviva

10x growth in mobile and table viewing in the last two years alone

27% of adults watch videos on non-TV devices every day

55% of people watch videos online every day

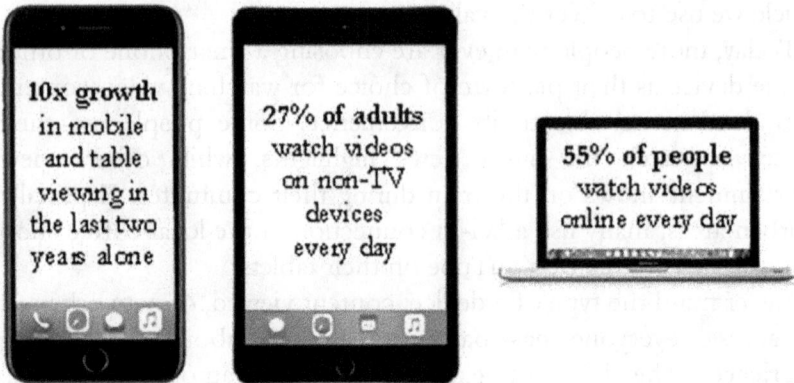

With QoE, the first place we need to start is to admit that we have a considerable challenge with understanding how our customers perceive the experience they receive from us. If you are in the 5% of telecom companies that already have solved this, then skip to the next chapter. Otherwise, we need to start working on defining the agreed QoE metrics from an outside-in perspective.

Depending on whether you are a single, multi or quad-player, you will have to determine the most relevant customer-centric QoE metrics for these respective business areas. It has to take into account the hardware, software, frameworks and platforms you are using. Then the hard work begins … and it is time to set in place robust measuring and insight solutions. You have to understand what kind of matters are issues for your customers and to what extent. Let's take an example from Conviva's Binge Watching Report:

- **Customer engagement with video content reduces by 14 minutes with just 1% increase in their buffering time**
- **That 1% buffering time will decrease the likelihood to watch content again from that same provider by 86%**
- **The impatience of customers on buffering time goes up by average 9 minutes every year, escalating the problems caused by poor Quality of Experience**

Now that we know poor QoE is the problem to acknowledge, we need to focus on taking into account every element that contributes to a customer's perceived quality of a service. This includes any relevant system, human, and contextual factors that contribute to that service. Reiter et al.[93] have identified the most important factors influencing the Quality of Experience and categorised them as follows:

Human Influence Factors

- **Low-level processing (e.g. visual and auditory acuity, gender, age, mood)**
- **Higher-level processing (e.g. cognitive processes, socio-cultural and economic background, expectations, needs and goals, other personality traits)**

System Influence Factors

- **Content-related**
- **Media-related (e.g. encoding, resolution, sample rate)**
- **Network-related (e.g. bandwidth, delay, jitter)**
- **Device-related (e.g. screen resolution, display size)**

Context Influence Factors

- **Physical (e.g. location and space)**
- **Temporal (e.g. time of day, frequency of use)**
- **Social (e.g. inter-personal relations during experience)**
- **Economic**
- **Task (e.g. multitasking, interruptions, task type)**
- **Technical and information (relationship between systems)**

How each of the above QoE categories manifests themselves in your business depends on what kind of services and customers you have. For example, to measure human influence factors, the Mean Opinion Score (MOS) is a widely used to measure the quality of media. It is a restricted view of QoE measurement (as most of them are individual), relating to a specific media type, in a managed environment and without explicitly taking into account customer expectations.

Traditionally, the MOS as an indicator of QoE has been used for audio communication, as well as for the assessment of the quality of

---

[93] "Influencing Quality of Experience" by Reiter, Ulrich; Brunnström, Kjell; Moor, Katrien De; et al

online video and television signals. Due to inherent limitations in measuring QoE with a single scalar value, the usefulness of the MOS is often questioned, but we know better than that. To meaningfully measure the QoE, we need to use a combination of measurements that make sense for both customers and the business. Let's take a couple of examples of such aggregate metrics for QoE.

Openet has created a Customer Experience Index (CEI) that combines relevant metrics from the network, QoE, VOC and other sources with weighted values. This kind of approach has the upside of balancing the customer experience measurements with different data sources into a single number. Naturally, the point is not the number itself, but the trend it takes over time. Here is an example of such a model, adapted and expanded from Openet's CEI metric[94] with other metrics we have discussed earlier in this book:

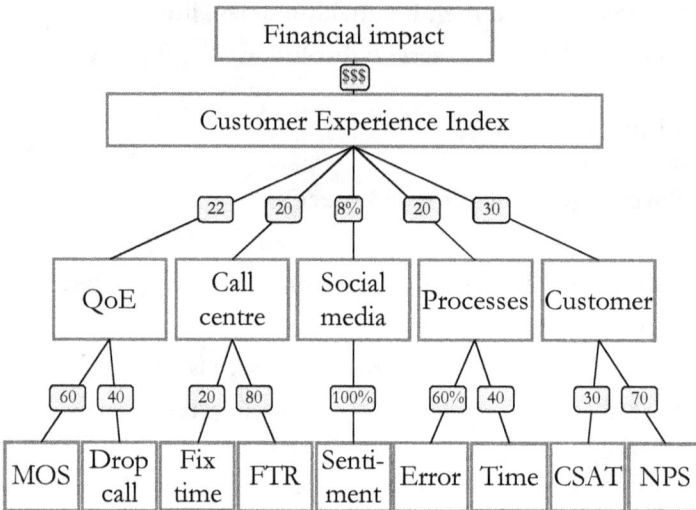

The above metrics and weightings are for illustrative purposes only. They show a model for creating a unique Customer Experience Index for your business.

Let's take another, more straightforward, QoE aggregate metric example from STL Partners' Mobile Network Experience Index (MobiNEX)[95]. It ranks mobile network operators by key technical

---

[94] https://www.openet.com/blog/customer-experience-index-%E2%80%93-nps-digital-age

[95] https://stlpartners.com/tag/mobinex/

measures relating to Quality of Experience. MobiNEX benchmarks mobile operators' network speed and reliability, allowing individual operators to see how they are performing compared the competition in an objective and quantitative manner from a technical perspective.

Each service provider is given a MobiNEX score out of 100 based on their performance across four measures that STL Partners believes to be primary drivers of customer experience: download speed, average latency, error rate and latency consistency. The thresholds they use are:

	ERROR RATE	LATENCY CONSISTENCY	DOWNLOAD SPEED	AVERAGE LATENCY
Weight	25%	25%	25%	25%
Raw Data Used	Error rate per 10,000 requests	Requests with total roundtrip latency over 500ms	Weighted average download speed in Mbps	Average total roundtrip latency in milliseconds
Top performance benchmark	25 per 10,000	1.7%	20.0 Mbps	142 ms
Low performance benchmark	100 per 10,000	50%	2.0 Mbps	500 ms

STL Partners use the above parameters to benchmark hundreds of operators around the world to get their MobiNEX metric. While this metric is highly technical and will not give a full view of Quality of Experience, it is still an excellent source of insight. If they benchmark your company, it is worth having a look. Perhaps you could even include your score as one of the CEI model metrics with an appropriate weighting?

"MARRIAGE" ADVICE BY DAVID MCGLEW, DIRECTOR OPENET

While it used to be that free Wi-Fi was enough to entice customers into a particular coffee shop or a hotel, this is no longer the case. Customers are starting to differentiate and are looking around for free high-speed Wi-Fi. This is not lost on the digital service providers who are rolling out high-speed carrier-grade Wi-Fi for their customers. As with the mobile and fixed broadband market, we'll see increasing use of superlatives in front of 'high-speed' to differentiate one Wi-Fi service over another: "Super high-speed", "ultra-high-speed" and so on.

Regardless of the marketing promises on the speed, if a service is slow due to congestion, customers will move on. Offering Wi-Fi and wishing that it will cope with the usage peaks is no longer good enough. Service providers need to ensure QoE to get customers using Wi-Fi as an extension to, and an alternative to, cellular. Due to the relatively low connectivity costs, Wi-Fi is becoming an extension to, as well as an alternative to, cellular networks.

We see the emergence of Wi-Fi first MVNOs, the rollout of multi-country Wi-Fi access points by service providers (including mobile operators, cable companies and fixed broadband providers) with Wi-Fi partners offering seamless authentication and access without any customer intervention. This is blurring the lines between Wi-Fi and cellular for many customers. It's network connectivity that enables customers to watch Netflix, connect with friends on Facebook and listen to music on Spotify.

Taking price out of the equation, as long as the quality is excellent and access instantaneous, most customers don't really care about the network. They just expect it to work and this includes delivering the QoE that customers expect. To deliver the optimum customer network experience on Wi-Fi, communication service providers need the tools in place that proactively manages QoE by setting rules for Wi-Fi offload from cellular and vice versa. Having Wi-Fi access bundled into mobile data bundles is becoming increasingly common. In Singapore, SingTel offers its Wi-Fi service bundled in with its mobile data plans. They claim that this is Asia's first Wi-Fi integrated mobile plan. The main benefits that SingTel promotes are:

- **Ease of use – Switch automatically between 3G/4G and premium SingTel network without a manual password login**
- **Wi-Fi Speed – Surf 5x faster than regular Wi-Fi services, at a typical speed of 4-10Mbps**
- **Coverage – Available at more than 700 hotspots nationwide, including popular shopping malls and busy MRT stations**
- **Low cost – Unlimited data usage at SingTel Wi-Fi hotspots**

The fact that network handover is automatic would suggest that, unless they check, users are unaware when they switch from 3G/4G to Wi-Fi. This opens up new opportunities for service providers to supplement cellular networks with Wi-Fi – assuming that the quality delivered is what the customer expects, and not worse than the 3G/4G service that they're used to.

Quality is now centre stage on Wi-Fi. But how can devices intelligently and automatically connect with the Wi-Fi access point that will deliver the right QoE for individual clients? Access Network Discovery and Selection Function (ANDSF) enables ranking of Wi-Fi networks under dynamic conditions, such as co-ordinating attachment to one network versus another when bandwidth is better, or network congestion is occurring on one of the networks considered.

ANDSF enables service provider controlled offload by helping devices to discover access networks in their vicinity (e.g. Wi-Fi) and make available rules to prioritise and manage the connection to all networks. This allows operators to control and define preferences dynamically – that is how, where, when and for what purpose a device can use a specific radio access technology – e.g. under what conditions traffic moves from cellular to Wi-Fi.

Service providers can define policies that enable devices to connect and authenticate to Wi-Fi access points automatically. From the users' perspective, this is an entirely seamless and transparent experience. The operator can centrally manage policies which give a great deal of control over which Wi-Fi networks will be chosen and under what conditions.

These network selection decisions can be based on multiple inputs including customer profile, historical data consumption, tariff plan, device type, location information, time-of-day, and a wealth of other network information. ANDSF enables operators' policies to be installed on users' devices and also to change them dynamically as conditions change. Do you already have such a capability in place?

III. A USER MANUAL FOR ALIGNING THE TECHNOLOGY TO SUCCESSFUL OUTCOMES

"You've got to start with the customer experience
and work back toward the technology,
not the other way around."
- Steve Jobs, Founder, Apple

WHERE SHOULD WE AIM TO GET WITH THE TECHNOLOGY?

"Know what your customers want most and
what your company does best.
Focus on where those two meet."
- Kevin Stirtz, Author, More Loyal Customers

If you wanted me to be cheeky, I could just say that *"the purpose of technology is to aid companies to deliver sustainable value and for customers to consume it"*. It doesn't have to be more complicated than understanding how customer experiences and technologies come together to create a digital telco, which serves customers better than ever while maintaining a healthy and sustainable financial situation for the business.

According to Capgemini's research[96], 58% of customers are willing to change over to a digital-only mobile operator if that was a viable option for them. That is if those operators would have available competitive data plans and well-performing online customer service. Capgemini's research also discovered that only 36% of consumers believe that their current mobile operator has used digital technologies to improve customer experiences. This reminds us of the study Bain & Co did already back in 2004, asking over 300 companies whether they deliver exceptional customer experiences.

80% of those firms claimed to do so, but when asked from their customers, only 8% agreed to that. The similar kind delivery gap seems to apply today on digital telcos, as consumers can see through the half-digitalised services that operators are trying to pass along. Many telecom processes start digital, perhaps by filling in an online form. But then most work after that first digital step becomes manual, completed in the background with a high likelihood of

---

[96] Unlocking Customer Satisfaction: Why Digital Holds the Key for Telcos, Capgemini

errors. For example, if you ever tried to change the ownership of your plan with Vodafone UK, you know what I mean. They have a fancy online form to fill in on their website. But after you submit the form, the digital process turns into a manual one and falls apart.

It may take Vodafone months, and you multiple calls, to get it sorted eventually. Every time you contact them you will have to explain the current situation from the start to the agent (who never reads the account notes to be up-to-date), and typically they have made several mistakes along the way you need to get them to fix. This happens because they have added 'smoke and mirrors' on their website to give an impression that it is a digital service when in reality it is not.

Why are digital customer experiences so crucial for telcos then? Let's take examples of statistics that will help you to justify the reasons for other leaders in your business:

- **80% of customers expect the digital experience to be at least as good as the in-store experience while 47% expect the digital experience to be better**
- **Mobiles are overtaking tablets, desktops and stores as the main channel of interacting with service providers**
- **74% of online consumers get frustrated with websites and content (e.g. offers and instructions) that are not relevant to them or personalised**

The above numbers speak a persuasive language, saying that becoming a digital telco is not a *"nice thing to have"* anymore, but a necessary business strategy that will ensure sustained success. Those service providers who will not embrace the digital side of the business will find themselves in deep trouble in the next few years to come. That challenge, however, is a big one to overcome due to complex technological architectures that are not designed for change and customer-centricity. Demand for structural requirements that come from the business and from customers, affecting the enterprise domains, are highly complex to implement in a fragmented and 'silo mentality' enforced architecture most telecom companies today have:

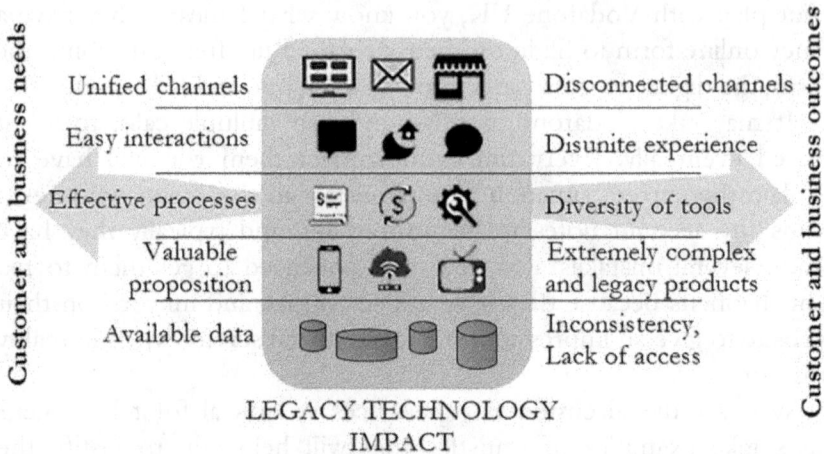

LEGACY TECHNOLOGY
IMPACT

As the above image shows, technology has the opportunity to aid or hinder all areas of business. Ultimately, for a digital telco, technology is the thing that makes or breaks the business as it enables a response to customer and business need and outcomes in the right way. The most common technology issues telecommunication companies have today are shown on the right-hand side. And these are the ones we need to address to be successful in creating remarkable digital customer experiences. To do that, there are three main areas that telecommunication companies need to focus on in their digital telco aspirations:

- **Redefine digital customer experiences**
- **Simplify operating models, and**
- **Build a digital customer-centric culture**

To redefine digital customer experiences, telcos need to focus much more on real-time, context-aware, mobile-first interactions. This means changing the perception of customer relationship from transactional to personal (and yet in mass-scale). Companies have to use the technology to extend their "human-touch" where it is not possible with humans due to scale or complexity. We'll talk about Digital Butlers, Quality of Network and other use cases that can create such emotions a bit later. As providing value through the network is typically the primary way for telecom operators to serve their customers, per-subscriber, pre-service view of network performance is really important for impacting customers positively.

Secondly, this needs to be achieved, with the help of technology, in a much simpler way than ever before. The times of building huge infrastructures are over. Today, technology needs to be reliable, cheap and agile. It merely has to contribute to simplifying value propositions, customer relationships and processes. Finally, all this has to come together through a digital, customer-centric mindset.

This is such a big challenge for traditional telcos, but as the Tele2 case study in this book has shown, it is possible, when done in the right way. The whole organisation has to be designed with digital customer experiences in mind, including community customer service, fully digital sales and delivery processes, digital value-based marketing and digital IT and network operations. The silo mentality that has kept business and technology groups from genuinely working together for customer value has to be removed throughout.

To aid you and your technology people on this journey, perhaps your organisation should set ground rules for your technology going forwards? The LeanCEM manifesto is a great place to start, especially for operations, so why don't we create a simple list of requirements that should apply to all your technology initiatives. Feel free to take this list and make it your own (as one size doesn't fit anyone), but I will give it go to get you started:

Our technology has to use all available assets (including, but not limited to other systems, data, inputs, etc.) to:

- **Sense context.** It needs to understand who is using it, where and what is the intended outcome of that interaction (whether with a customer or with an employee). This can be done either by combining various data sources or by direct input from the user.

- **Anticipate needs.** The technology should know as soon as possible what the user needs and proactively offer that.

- **Be human.** The technology works for us, not the other way around. The technology has to be able to adapt to each user's personal preferences, style, language, etc. to make it feel like they are interacting with another friendly person rather than a mindless system.

- **Self-aware.** The system should understand how it is performing if there are any issues and what the areas where it could be improved are. This means logging and reporting internal and external performance needs to improve over time.
- **Be connected.** The technology needs to have access and be able to utilise any assets and data that are available.
- **Be as safe as Fort Knox**[97]. Protect against threats seamlessly, transparently and effortlessly.
- **Be valuable.** Automate everything and anything that is possible. Minimise any manual labour. Save human resources for meaningful work.

The above list could get much longer, but as agreed earlier, this is just to get you started. If you want to have a small play around, take the system that you use the most (or your customers use the most) and think about what that system would look like, if it followed the above principles. Would it be much better than what it is today? I bet so.

Let's take, for example, Business and Operational Support Systems (B/OSS) and put them through this lens. Historically, B/OSS solutions have matured from back-end systems to be an integral contributor to Customer Experience Management. Nowadays, these systems are crucial tools in a digital service provider's network management and customer service. From OSS perspective, such systems support network operations with testing, probes, agents, fault reporting, issue alarms, service management and assurance, as well as activation and provisioning services.

On the other hand, BSS focuses on dealing with customer service processes, invoicing services, security and fraud management. B/OSS solutions are used extensively by operators to manage complex networks and IT infrastructure, service delivery, and customer support. Thus, investing in B/OSS improvements are vital to delivering great digital customer experiences going forward.

For B/OSS systems to be able to fulfil the expectations described

---

[97] https://en.wikipedia.org/wiki/United_States_Bullion_Depository

earlier, it requires combining customer and network information traditionally managed in the CRM system with proactive capabilities around network, and service assurance that can detect network problems impacting customer services and take proactive and pre-emptive measures to counter.

Such requirement also implies a well-justified need for pre-integration of historically separate OSS and BSS functions. Such an approach will allow digital service providers to ensure that even low-level technical processes contribute to improved customer experience. An integration of systems and data sources provides a foundation for the 360-degree view into the customer experience and gives insights for telecommunication companies to act upon.

As we already know, Customer Experience Management initiatives can be successful only when a telco achieves a holistic view of the customer. Leveraging the tools mentioned above from the CEM arsenal can help telcos differentiate their offerings in the minds of the customer. B/OSS systems need to change a great deal from what they used to be, to now being able to support communication service providers to have an end-to-end view of the customer experiences enabled by the digital technologies, IT and network.

THE FOCUS RECALIBRATION

"If we want to know what a business is, we have to start with its purpose. And the purpose must lie outside the business itself. There is only one valid definition of business purpose: to create a customer. The customer is a foundation of a business and keeps it in existence. The customer alone gives employment."
- Peter F. Drucker, Social Ecologist

I wish everyone would read Peter Drucker's management books. Not only do they cure insomnia, but they also give ideas that are paramount for a successful business. They are simple and essential concepts written in a complicated language like in the quote above. That's why it's a good thing that some people have deciphered what he means and expressed it in plain English. This same goes for technology! We tend to overcomplicate things when it comes to solving everyday problems with high-end tech. That is because we are not doing what Steve Jobs suggested when he said to start with the customer in mind and not the technology.

Do you remember the cute cats from the 'A CEM FRAMEWORK? WHAT'S THAT?' chapter? You may have noticed that there was a grey background behind the elements introduces as part of the CEM framework. That was to represent not only the strategic alignment but also operational one. The grey colour in that picture represents how everything in the organisation has to come together to deliver successful customer outcomes. And using technology to do that as part of operations makes more than sense, it's necessary.

Sure, the technology was a separate box at the bottom for drawing technical reasons, but we need to keep in mind that technology has its *"tentacles"* spread everywhere. And that is also why it so often gets more power than it deserves. Technology is here to serve us, not the other way around (unless artificial intelligence becomes smarter than we are and take over the world?). Why don't we go back to that picture and draw it from a technological perspective?

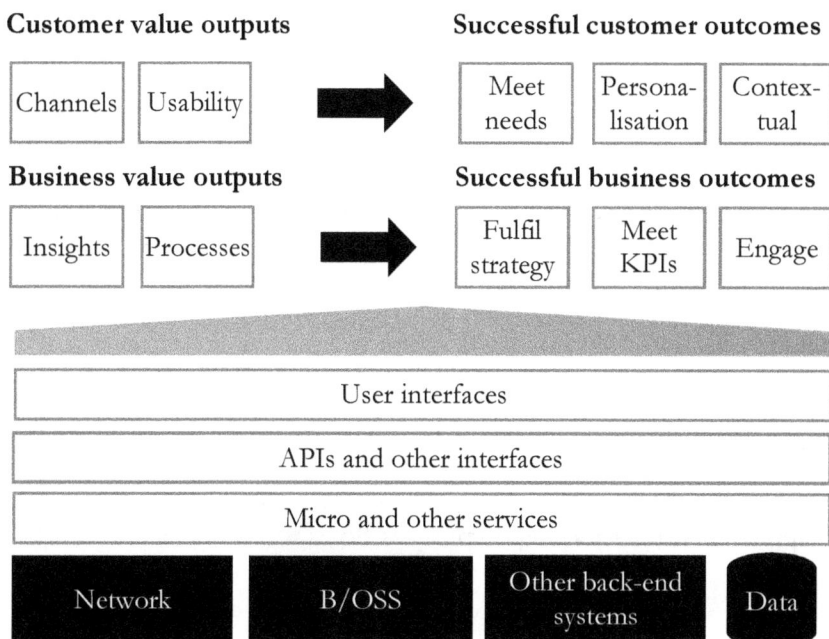

Customer value outputs

Channels	Usability

Successful customer outcomes

Meet needs	Persona-lisation	Contex-tual

Business value outputs

Insights	Processes

Successful business outcomes

Fulfil strategy	Meet KPIs	Engage

User interfaces

APIs and other interfaces

Micro and other services

Network	B/OSS	Other back-end systems	Data

The above picture could take very different shapes depending on how your business has dealt with technology so far. What should be common though is that the technology contributes to business customer outcomes through specific outputs that are understood. Too many telecommunication service providers have lost themselves in the technological jungle as they have kept on adding more and more systems and solutions, without keeping a map like above updated. Such lack of discipline has led to misalignment between what the company wants to produce (e.g. customer experiences) and what it actually delivers (e.g. customer perceptions of those experiences).

Completing a comprehensive review of the above alignment takes discipline. Each layer needs to be connected to each other to reveal what is going on in there; almost like peeling an onion. Or getting to know your new girl/boyfriend. Once you understand one layer, you can move on to the other one, but then you also need to go back to see how the interactions between these layers work. In practical terms, you can align your business and customer experience strategy together using the previously mentioned Customer Experience Blueprint technique.

Then you can use the Customer Journey Mapping technique together with Breakpoints and Business rules to see how that layer connects to your business processes. Then you can use BPMN modelling technique to map processes and technology together while utilising Enterprise Architecture tools. Sure, none of these tools are new! But using them all together to create an end-to-end, top-to-bottom transparency is still quite rare. And yet so powerful. I am sure your business has at least part of the elements described above, but do you have visibility into the interactions between these layers? Probably not.

Let's take a practical example from a European telecommunications operator who has done this. To protect their business I am not able to disclose all details, but it is enough for us to see how this has worked exceptionally well for them. Let's start by mapping customer journey to process flows.

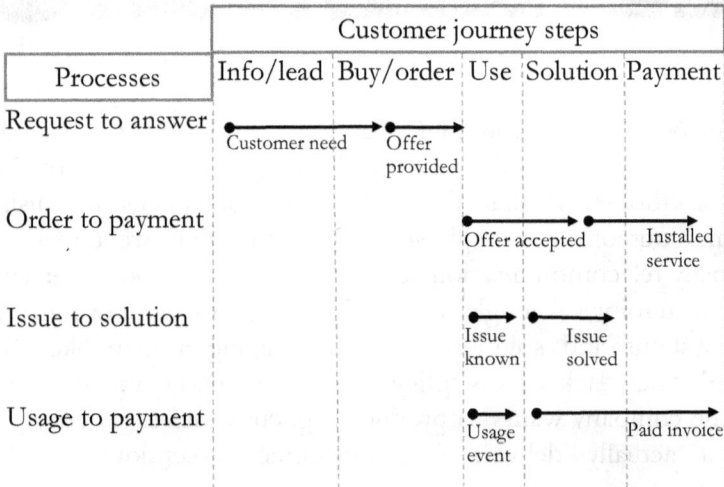

Processes	Customer journey steps				
	Info/lead	Buy/order	Use	Solution	Payment
Request to answer	●——→● Customer need · Offer provided				
Order to payment			●——→● Offer accepted		Installed service
Issue to solution			●→● Issue known · Issue solved		
Usage to payment			●→● Usage event		Paid invoice

The above image illustrates a simple way of mapping customer journeys and business processes together. It does not have to be more complicated than this to get started. You can then evolve your approach, depending what is useful in your company particularly. Next, what you could do is to map the processes to systems using the same approach. It could look something like this:

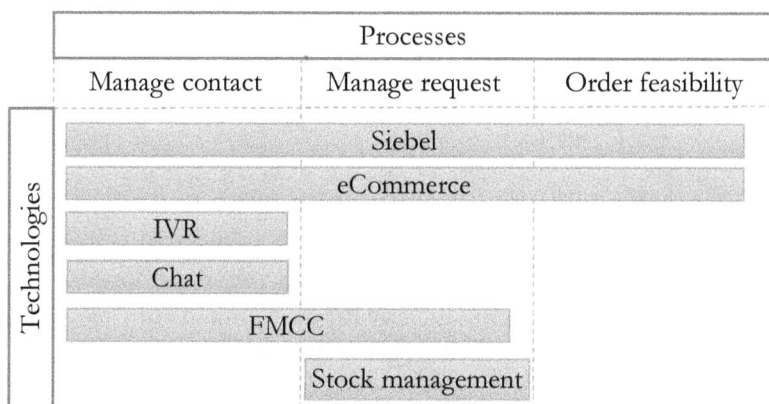

Processes		
Manage contact	Manage request	Order feasibility
Siebel		
eCommerce		
IVR		
Chat		
FMCC		
	Stock management	

Just having something like the above chart will help your business to understand how different processes and technologies are linked. Now you would have a comprehensive, end-to-end model from all the way top to the bottom, including customer and business strategies, customer journeys, business processes and technology. Based on my experience in dozens of projects, it does not take much time to get the first draft, and it will give impactful insights straight away into where the gaps and issues are.

The focus recalibration means that your business will not see the technology anymore as the end, but as a means to the end. And that end is wherever the company can and wants to add the most value to its customers in a sustainable way. Regardless of what your role in the company is, it is your responsibility to keep this in mind always and intelligently use the technology. In next chapter, we can get our hands really dirty and start going through exemplary use cases (i.e. solutions) that will drive different customer outcomes while taking into account all the principles that have been discussed so far.

CUSTOMER-CENTRIC USE CASES FOR TELECOMS

---

*"The telephone has too many shortcomings to be
seriously considered as a means of communication.
The device is inherently of no value to us."
- Western Union, Internal Memo, 1876*

L ack of vision is a horrible way to die. When I first started to work in the telecoms sector, I was surprised by how much of the business direction was driven by existing use cases and solutions. It is easy to see how that has come about in an industry that historically has been so heavily technology driven. A mentality of *"since someone else has done it, we can do it, too"* has driven much of what happens in the industry. Today, it seems that those days are gone. Buying technology that enables a limited number of use cases is dangerous. It is like only ordering brochures on outdoor weddings, while you won't know if it will rain or shine on that day.

You want to be able to adapt to whatever the weather will be on the day and be prepared for both if need be. So it is with the technology, too. It needs to be able to adapt to changing market and customer needs, wants and expectations. Thus, the use cases discussed in this chapter are not to be only or even the best use cases available. They are just here to give ideas from a customer-centric perspective. You will have to be able to make them your own rather than just take them.

And that's ok. You have enough tools from this book and from 'The 5-Star Customer Experience' to do that in a way that will ensure both customer and business success. Each of the following 12 areas covers something that is valuable and interesting to the customers. Several of them contain many smaller use cases to address a customer or business need and some just one. The use cases we will cover are:

- **Happy Happy Joy Joy Customers**

- **Let's Have a Chat - Channel To Channel**
- **Let's Make The Almighty Dollar To Come Alive**
- **Yes, Of Course, I'm listening**
- **Will You Marry Me?**
- **Have Your Own Digital Butler**
- **Look What I Found, You'll Want One, Too!**
- **What a Pleasant Experience This Has Been**
- **Squeeze the Last Drop Out Of the OPEX**
- **Customer Experience Based Charging**
- **Get Rid of Those Bloody SIM Cards Already!**
- **Please, Don't Leave Me!**

They focus on essential areas of Customer Experience Management, such as customer satisfaction, churn, Net Promoter Score, and many others. Let's start with the first one that is about customer satisfaction.

Happy Happy Joy Joy Customers

"When a customer complains he is doing you a special favour; he is giving you another chance to serve him to his satisfaction. You will appreciate the importance of this opportunity when you consider that the customer's alternative option was to desert you for a competitor."
- Seymour H. Fine, Author, Social Marketing[98]

It is one thing to put a Customer Satisfaction (CSAT) programme in place, but using it to deliver consistently positive interactions with customers is much more challenging. Nonetheless, it is likely that CSAT will remain strategically important to communication service providers for many years, as evidence shows a strong correlation between CSAT and revenues[99].

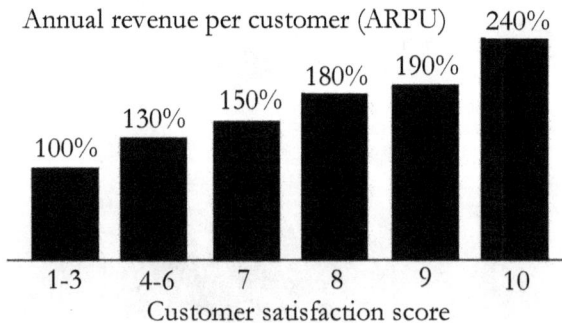

Annual revenue per customer (ARPU)

	100%	130%	150%	180%	190%	240%

Customer satisfaction score: 1-3, 4-6, 7, 8, 9, 10

Other research shows that customer satisfaction is closely aligned with customer's loyalty and willingness to trust and recommend their

---

[98] Source: https://www.salesforce.com/hub/service/famous-customer-service-quotes/
[99] Source: Medallia research

service provider[100]. All of these are vital contributors to customer lifetime value and market share. Telecom companies that focus only on new customer acquisition will struggle to meet their business targets as they will also need to replace unsatisfied customers leaving the company – so it makes clear business sense to keep existing customers satisfied.

And how will customers evaluate whether they are happy or not? They will ask themselves the question: *"Is it worth it and easy enough to do business with this company?"* and if the answer is yes, they'll stay in business with you. The secret to improving customer satisfaction is first to understand what customers want, what they need, what they expect, and then to offer just a little bit more than that in the right areas[101].

You don't need to wildly exceed their expectations (there is actually research showing that it is a waste of your money), but you need to be just a little bit better than expected. And how would you know what they expect? That's where the previously discussed Blueprint, Journey Mapping and Voice of Customer techniques come to your aid. Now, let's go through the use cases that might give you ideas on how to improve customer satisfaction:

It's Not Just About You

Customers are used to thinking about themselves when it comes to telecommunication services – but we can change that. Allowing customers to donate their unused balances to their favourite charities is a win-win - charities can spend more of their budgets on doing good, and customers can feel better about themselves by making that possible.

Rolling Refunds

Billing may not seem very sexy as a differentiator and a way to reduce churn, but Google is managing it. One of the most innovative elements of their Project Fi offering is providing data refunds and making no extra overage charges. For example, if you pay $40 for 4GB of data in a month, but only use 2.5GB, you will be refunded $15 for the 1.5GB you haven't used – and this will come up as a credit on your next bill as a top line item. This may seem a little like a

---

[100] Source: UK Customer Satisfaction Index, 2016 report

[101] Or alternatively watch this educational video from YouTube http://bit.ly/happyhjj that explains the title of this chapter

'rollover data' use case, but it is dollars and cents to the customer rather than GB in their pocket.

Tenure-Based Grace Periods

Even now, many operators' first instinct is to block or throttle their customers when they reach their limits - undesirable outcomes for both operator and customer. Offering grace usage amounts based on tenure – how long they had been subscribers - would be a very customer-centric move. For example, customers with one year of tenure could get a 10% grace amount on their data bundle so 1GB bundle customers would get 100Mb to play with if they reach their limit – and the percentage would increase with length of tenure.

Data Gifting

Many subscribers regularly consume less than 50% of their allowance, presenting an opportunity for the provider to offer 'data gifting' to its subscribers. Etisalat in the UAE has introduced a service that lets customers send a data bundle to anyone on the same network. As an extra gift from the operator, when one subscriber gives a data gift to another, Etisalat adds a 10% bonus. This has the dual benefit of boosting customer satisfaction for the benefactor and encouraging data usage for the receiving person. Glo in Nigeria and Globe in the Philippines are examples of operators that have successfully introduced similar schemes.

Usage Notifications

Trust is one of the key drivers of customer satisfaction and loyalty in the telecom industry. Key to establishing trust is charging customers accurately, with no nasty shocks at the end of the month. Putting real-time notifications in place for customers about to exceed their limit helps to maintain customer awareness. This might sound like a routine 'hygiene factor', but the list of operators who don't already have it in place is surprising.

Comcast, for example, trialled a $10 charge for customers in certain markets who exceeded their 300GB limit and it didn't go well. Customers started tracking their own usage and found it did not match what Comcast was telling them. Comcast later shared that they had only 94% accuracy in usage charging and shortly after that abandoned the trial. Winning back customers' trust after such an event can take years.

Abolish Overage Charging

Operators for many years have been wedded to overage charging on customer bundles as it gives them a welcome short-term revenue bump – but it has a less pleasing impact on customer satisfaction and lifetime value. A better customer experience would be realised, by, for example, bumping over-spending subscribers down to 2G speeds (and zero-rating their traffic). Customers retain minimal access to data, can purchase an additional data pass if they want to improve it, and the cost to the operator is minimum.

Allow Device Tethering

If an operator has real-time data usage tracking, advanced notifications and bill shock control in place, it is no good to block device tethering (i.e. using the handset to stream data to another device). It's more efficient and effective to monetise tethering via smartphone charges – not least because customers who utilise their entire data allowance are likely to buy additional data passes and boost operator revenue.

No Best-Before Date for Data

Most operators restrict rollover of data, missing a monetisation opportunity. Here's one idea: Operator X sets up a 'rollover club' which, for a $1 subscription per month allows customers with unused data to extend its validity for up to a year. This differentiates from operators allowing no, or very limited, roll-over and creates an attractive 'community' message.

Let's Have a Chat - Channel To Channel

Omni-channel is a tricky Customer Experience Management issue. There are not many customers who actively think about channel switching, they just want to use whichever channel is most favourable for them at any particular time. Customers only switch channels if they can't do what they want in the most convenient way in the channel of their original choice.

And in the context of customer service, clients do not desire Omni-channel when dealing with problems related to operators; they want the problem solved for once and for good in any channel they prefer. And yet, in reality, an Omni-channel experience looks something like this for the customers:

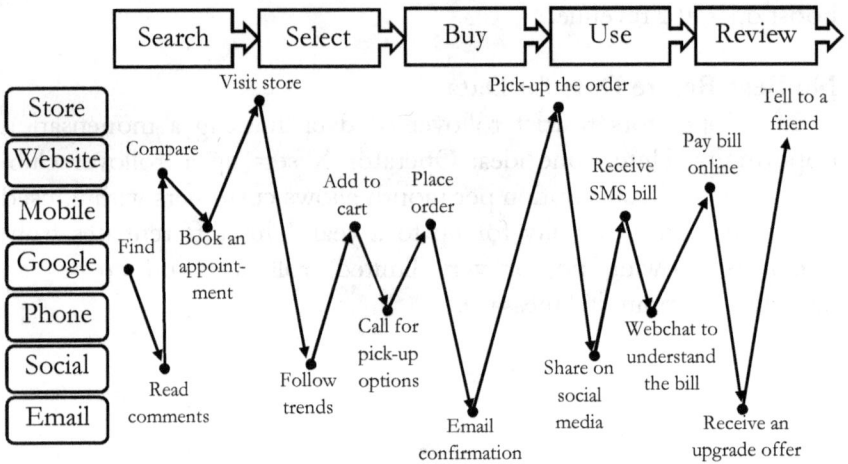

If your company has already conducted a customer journey mapping exercise, then you should have a similar map available. It will show you how many times the customers would potentially need to switch channels to get the task completed. In a perfect Omni-channel world, those switches happen seamlessly and without errors. In an ideal customer experience world, the customer would get the job done in any one channel they choose to, without a need for extensive channel switching.

Where we have come today is that 80% of the consumers use three or more channels to reach their operator (Source: Optymyze) and yet only one in five of those service providers are able to have contextualised discussions with their clients across all available channels. That is despite the fact that 74% of communication service providers believe that contextual discussions are highly impactful (Source: Openet & EuroComms study in 2017[102]). McKinsey & company make bold claims on benefits of delivering exceptional Omni-channel experiences[103]:

- **+36% on customer satisfaction**
- **+19% on likelihood to renew contract**
- **+28% on willingness to recommend (NPS)**
- **+33% less likely to churn**

Naturally, the benefits your business might receive from Omni-channel initiatives may be different, and certainly worth the look. To maximise your potential to be successful in Omni-channel projects, following these principles will take you far:

- **Put the customer at centre of your Omni-channel strategy, and align it with Customer Experience Blueprint, journeys and Voice of Customer feedback**
- **Establish a governance model for the Omni-channel programme**
- **Remove silo mentality and unify data access**
- **Invest in simple, agile technology that will change with your and customers' needs**
- **Improve continuously, set a baseline and targets for improvement and ensure there is continued action to make your Omni-channel results better**

Finally, let's also take examples of use cases that could help your Omni-channel efforts:

Advanced Notification Channels

Traditional channels like email, SMS, and USSD are being augmented by Google and Apple push notification services on smartphones, including in-browser notifications from Google Chrome. The World Wide Web Consortium is working on a push API, which will allow operators to send usage notifications to customers even if their self-care portal is inactive.

Variable Offer Presentation

Omni-channel presence is essential for customers who should be able to see the same offers presented the same way across all channels representing the Service Provider – or differently if the service provider chooses. Service Providers can use this capability to promote the use of one channel over another – or can basically provide more options to the customer and let them choose.

Connected Car

According to a survey of 2,000 new car buyers by McKinsey, 13% will no longer even consider a new car without Internet access, and more than a quarter already prioritise connectivity over features such as engine power and fuel efficiency. In 2015, AT&T announced a partnership with Audi to provide LTE data connectivity SIMs to the Connected Car platform for their entire Audi range.

AT&T currently offers two connected car packages for Audi owners: a 1GB plan for $10/month or an unlimited plan for $20/month. The connected car platform provides services such as navigation, streaming and high-speed internet access for up-to-the-minute traffic information, route guidance, over the air map updates, internet radio, social media and personalised RSS news feed.

Help Anytime, Anywhere

When customers are presented with a problem, they want speedy resolution. These days the first attempt is often online, through FAQs, online support, videos and helpdesk groups. If the answer is not found, then they turn to CSRs through e-chat, call centres or physical stores. Because the channels are not connected and synced in real time, they often, find themselves going over and over the same details. Enabling customer information in real time in every channel is key to providing adequate support for customer problems.

Let's Make the Almighty Dollar to Come Alive

---

*"All business success rests on something labelled
a sale, which at least momentarily
weds company and customer."*
- Tom Peters, an American Writer

Customer Experience Management initiatives are becoming ever-more sophisticated and challenging to implement. The number of investments made in this category will only increase. Having a full understanding of customer expectations and responding to those appropriately will ensure the business continues to thrive. That is why every interaction a service provider has with its customer matters.

Organisations must focus on helping customers achieve their goals through those interactions. Not least as highly-engaged customers deliver a 23% increase in profitability and revenue over the average customer (Source: Harvard Business Review) and businesses with higher levels of engaged customers outperform peers by 26% in gross margins and 85% in sales growth (Source: Bain and Company). Let's go through examples of use cases that impact revenues and customer experience positively.

Digital Asset Trading

Trading has become a favourite hobby for many urban people. Why not let them trade their digital assets, too? For instance, someone with a considerable number of voice minutes left in their plan could swap them with another subscriber for more data.

Data Gifts

Offering customer bundles of free data is a great way to publicise a new service launch. T-Mobile Netherlands utilised this a few years ago with significant impact, giving all new and existing customers 4GB of data to promote their nationwide 4G launch. Customers prompted the freebie on social media; T-Mobile added a nice twist to the offer by allowing customers to gift their data to a friend or family member.

One Size Fits No One

Service providers need to be flexible enough to accommodate widely varying customer situations. Customer insight allows operators to design and implement data passes for specific contexts and situations, such as social media use, time periods, exceptional download needs (such as a movie) or location – for roaming, for example. These will give customers a greater sense of a well-targeted proposition.

Honey Trap for BYOD

'Bring Your Own Device' (BYOD) offers value to the enterprise and the user. If an operator can establish a VPN tunnel to separate corporate and personal usage, the employee can reclaim the work element of their mobile usage without contention, providing cost savings for many enterprises. Operators who successfully deploy this functionality could see a flood of new enterprise customers.

Retail Wi-Fi with Tracking

Retail and telecommunication businesses can help each other. Imagine a situation where a customer enters a store and gets free access to high-speed Wi-Fi. Great value for the customer, and even better value for the retailer, who can use this technology to see when and where the customer is in the store, with accuracy all down to the aisle and even product level, giving valuable customer insight.

Send Offers that Matter

Customers have little patience with unwanted messages and will often turn off push notifications. However, 48% of customers surveyed by Localytics said they want offers that are relevant - for example, a customer using all their data allowance on YouTube is likely to be interested in an "All the YouTube you can eat" offer. Other ideas for context-specific offers for your customers might include Bill Cycle dependent upgrades, Data Push and Pull opt-in/opt-outs, or up-sells to larger data bundles.

Entertain Me

Consumers love to play games - one-third of consumer's time on mobile apps is spent on gaming and entertainment (Source: Flurry Analytics, 2016). Consider how you can entertain your customers along their journey and tie it to your brand. Games can be fast and straightforward, for instance, scratch-to-win or spin-the-wheel, or more profound, perhaps enhanced with image recognition or augmented reality. You can improve the experience by making it social, allowing customers to share the experience with their friends. Keeping users engaged and entertained will keep them coming back to the mobile app – where they might happen to see a new offer...

Content Partnerships

UK operator Vodafone has led the way in content partnerships which encourage people to use their LTE service. In 2015, for example, they included a 12 month free Netflix account when customers signed up for their fixed line broadband service. Operators could extend similar partnership offers to create a one-stop shop for customers - free Netflix for mobile and fixed broadband, for example.

Yes, Of Course, I'm listening

---

*"The most important thing in communication is
hearing what isn't said."
- Peter Drucker, Business Guru*[104]

L istening to customer feedback and combining it with other data
sources (from the network, for example), powered by advanced
predictive analytics, gives digital service providers the ability to
develop and refine Customer Experience Management initiatives with
far greater precision and impressive results. Insight-led solutions help
telecoms deliver a better experience to their customers, and to that
end, it's crucial that telcos invest in their customer listening and
interpretation capabilities.

The most advanced 'Voice of Customer' (VOC) programmes use
sophisticated analytics to make sense of previously untapped data
from multiple channels, harnessing sophisticated data analysis models
to help companies anticipate trends, make better decisions more
quickly, and differentiate from their competitors. That rewards them
with higher customer engagement that has a direct impact on the
bottom line results[105]:

---

[104] Source: https://www.brainyquote.com/quotes/quotes/p/peterdruck142500.html
[105] Source: PeopleMetrics

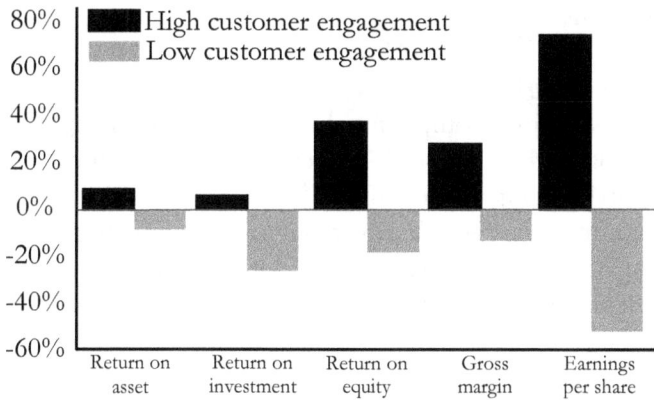

Insight-driven solutions powered by predictive analytics are a critical piece of Customer Experience Management. The right decisions are vital if a company wants to minimise churn and gain brand advocates. High levels of customer engagement improve many financial metrics, including return-on-assets, investment, and equity, as well as gross margin and earnings per share. None of these benefits will be realised unless customer insights are used to drive business action. Let's go through examples of use cases related to listening to customers and using that to their benefit.

Smart Pricing Alerts

Making customers' lives easier is an important part of successful Customer Experience Management. In 2017, Vodafone started to use customer information to create personalised and timely offers. 'Smart Pricing Alerts' offered customers additional data services depending on how they were using their device – such as a data add-on when the customer was using Wi-Fi whose performance was inferior to the available cellular coverage, or data top-ups when a customer reached 80 percent of their allowance.

Small Change

Throughout this document, we talk about how understanding customers is a necessity rather than an option. Here we see a great example of customer-centric innovation in practice, in a low-ARPU market. Vodafone Egypt's Fakka concept allows local stores to give prepaid vouchers instead of small change to consumers for their daily purchases. This boosted ARPU by 7% for Vodafone - a great example of an innovative proposition targeted at the right customers at the right time.

Advertising Insights

For years, cable companies have relied on sample audience viewership data from Nielsen or its equivalents. In a time where data is king, not knowing the details of their customers' viewing habits puts cable companies at a disadvantage over their data-rich and data-competent OTT competitors. Gaining detailed second-by-second measurement of their subscribers' activities opens up stacks of opportunities.

For example, data relating to an advertisement events and correlated content may be sold to advertisers to measure the success of their campaigns. If a cable operator knows if and when customers channel hop during ad breaks, they can sell this information to the TV networks - allowing them to gain a premium price for their advertising slots. Time Warner Communications (TWC) rolled out their Kernel Connect solution for advertisers in February 2016.

Content Negotiation

Access to content has long been a thorn in the side of satellite and cable providers. On average, programming costs swallow up 59% of video revenue, by far their most considerable operating expense. Viewership levels for television networks, especially in the US, are falling, but the price they achieve for their content continues to grow. One cable service provider in the United States saw video revenue fall 1.3% in Q2/2015 YoY but saw programming costs rise 11% in the same period.

Coming to the negotiating table without detailed analysis of viewership behaviour and patterns is like (as the saying goes) bringing a knife to a gunfight. If a cable company can identify poorly performing channels with low viewership, however, they can renegotiate carriage or even drop the service. Real-time viewing measurement puts cable providers in a far better negotiating position, making their content gambles a little bit safer.

Shared Data Plans

Different age groups use varying amounts of data, and the Australian operator, Optus, used this insight to be more innovative with their Shared Data family plans. Optus saw that around 68% of subscribers on My Plan were using less than 50% of their monthly data allowances. They also saw that 18-20-year-olds on the plan used almost three times as much data as those over 45 years of age.

They changed their Family Sharing offer, allowing subscribers to pool their data and maximise the value of their mobile plans, resulting in a superior customer experience for their subscribers. Optus gained additional devices and new subscribers to their data plans, boosting overall spend per customer, and increasing customer tie-in.

Multi-Device Plans

According to E-consultancy, more than 60% of online adults in the United Kingdom use two data devices every day and close to 25% use three devices or more. The number of devices that are internet-enabled is moving far beyond smartphones and cellular-enabled tablets. As more data-enabled devices come on stream, they create opportunities for new revenue. As the number of devices and average data bundle size increases so does average revenue per account (ARPA). For instance, Verizon has increased its device per account ratio from 2.4 to 2.8 since launching Shared Data plans, to which devices like cameras, game consoles and connected cars can be added. This is an excellent example of putting customer insights into business action by offering multi-device shared data.

Will You Marry Me?

Marriage must be the ultimate loyalty programme, and hence this book was also themed with marrying our customers. A study by the Chief Marketing Officer (CMO) Council found that 12% of customers switched from service providers who rewarded their loyalty poorly. And as we know, it can cost the average business five times more to acquire a new customer than to retain a current one. Loyalty incentives and exclusive membership programmes are now essential elements of Customer Experience Management in telecommunication industry where competition is brutal.

Whether you to choose an experience programme, membership programme or loyalty programme will depend on your Customer Experience Blueprint. Research conducted by Temkin Group in 2017, based on data acquired from over 10,000 consumers, showed that customer experience correlates closely with loyalty, emphasising that the best customer loyalty programmes improve the overall customer experience. So, let's explore further, how we could impact customer loyalty through example use cases.

All for One and One for All

The more your customers share, the faster you build revenue, so why limit loyalty rewards to individuals? Allow dynamic pooling of data allowances between family members, friends or other affinity groups. The more the group spends together, the more they earn cashback.

Mobile Birthday Offer

Who doesn't like a birthday gift? As insignificant as it might seem, receiving a gift on your birthday is always welcome. A Mobile Birthday Offer of 1GB on a first subscription birthday, or free Facebook access in your birthday month - thanking the customer for being part of your family can increase loyalty.

Find that Pokémon

To engage customers more with your brand, deliver digital rewards through real-world gameplay. As in the Pokémon game, customers can complete tasks and play games to gain more data, voice minutes or other rewards. The opportunities are vast, and the initiative can be targeted to receptive customer segments.

Points to Value

There are various points-based rewards programmes around the world (e.g. grocery store points and air miles for flying). These points are useless until they are redeemed and turned into real value. Why not make partnerships with existing points-based loyalty programmes and let their customers redeem points for your services and products? Or make it a two-way street and let your customers earn loyalty points on their preferred programme.

I Want It Now

Many people prefer instant gratification over long-term loyalty points collection. This requires you to have the data and systems needed to direct behaviour that is most fruitful for your customers and the business through instant rewards - such as 1 GB of free data after topping up a prepaid contract with 500 minutes.

Loyalty Data

Everybody loves free stuff, but rather than simple giveaways of mobile data why not link rewards to the subscriber's mobile anniversary. When a customer hits one year of tenure, the operator could provide them with a 1GB bonus for a one month period, helping them feel the love when they are about to reach the end of a 12-month contract. Follow it up with a 2GB data bonus on the second birthday, 3GB on the third birthday and so on. This offer has been a winner for KDDI's mobile brand Au in Japan.

Pay-on-Time Rewards

Mobile operators can reward prompt payers with a bonus. This not only reduces the operator's bad debt bill but also makes sure customers stick around longer. Boost Mobile, a sub-brand of Sprint in the US, has instituted this with great success. For a low cost, they've reduced churn and improved cash flow. Another approach would be to give three months of free Spotify to reward customers whose payment record has been enhanced over the past 12 months.

Loyalty Programs for Non-Customers

Three Ireland have an excellent rewards approach that boosts customer satisfaction and loyalty. All 3Plus customers get exclusive access to 3Arena and to Ireland's top music festivals, sports events, presale tickets and guest lists, as well as discounts and offers from top Irish brands. For a limited time in 2015, Three opened up access to this app to any Irish mobile customer even if they weren't on the Three network, creating a marketing channel for Three similar to the Swisscom Mobile TV example. This is an innovative way to begin a conversation with non-customers and gives Three's customers good reasons to recommend them to friends and peers.

Have Your Own Digital Butler

---

*"Biggest question: Isn't it really 'customer helping'
rather than customer service? And wouldn't you
deliver better service if you thought of it that way?"
- Jeffrey Gitomer, an American Author*

Did you know that half of your customer base prefers to solve product or service issues by themselves and 70% expect a telecommunication company to offer a self-service application (Source: SuperOffice). Gartner predicts that by the year 2020 customers will manage 85% of their business relationships without human interaction. Convenience and effort are a significant contributor to how a consumer views the service you offer. Investing in self-service tools across your platform is vital for customers that prefer not to queue for problem resolution. Let's go through use cases that help to enhance your self-service capabilities.

Create a Digital Butler

T-Mobile has found a funny way to serve their customers by creating a digital butler called Horst. It is a customer service Chatbot with a personality. For that, T-Mobile has won the 'Best Online Service' award both in 2016 and 2017 in Germany. Though creating Chatbots is still technologically very challenging, there are many telecommunication companies, including the Tele 2 interview in this book, who are looking to expand their customer service this way. Perhaps you should, too?

Make It Mobile

The easiest way to enhance self-service capability is to acquire a mobile app that has all the necessary essential functions giving the customer the control over their services. Implementing such will remove the need for customers to contact the call centre and save you money on OPEX. Combined with the personalised offering, predictive data and other vital features of CEM, you can also use it as a channel to up and cross-sell your services with a high-impact, low-cost approach.

Build Your Plan

Letting customers build their own plans is about more than cost saving. Unsurprisingly, customers are happier with plans that are well-suited to their needs. This also drives up advocacy as it is easier to recommend a perfectly matched service than a more approximate bundle.

Set Your Spend Limits

Allowing customers to set limits on their monthly spend via a mobile self-care app is a tried and tested way of giving cost-conscious customers certainty on their mobile bills and helping them avoid bill shock.

Look What I Found, You'll Want One, Too!

"Don't find customers for your products, find
products for your customers."
- Seth Godin, Marketing Guru

Net Promoter Score (NPS) remains the telecom industry standard for measuring customer advocacy (i.e. how likely it is that a client will recommend you to their friends). In 2017, TM Forum conducted a survey where they asked how many communication service providers use NPS as a KPI and 43% reported to do so. However, 57% of those same respondents said they don't know their company's Net Promoter Score, which is quite sad.

To successfully improve its Net Promoter Score, a telco has to move from being a score 'watcher' to being a player. This requires a robust 'Voice of Customer' programme with the strategic goal of improving Net Promoter Score through advanced analytics and action. This is increasingly important as the speed of new service adoption is increasing exponentially:

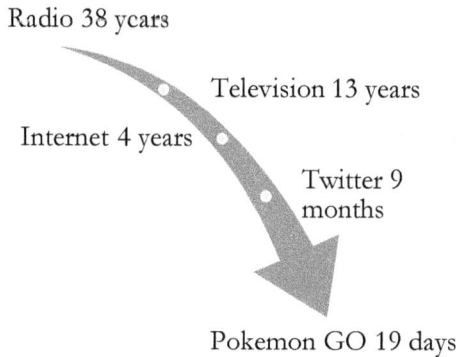

Radio 38 years

Television 13 years

Internet 4 years

Twitter 9 months

Pokemon GO 19 days

Such an impressive speed of adoption would not have been possible for Pokemon Go without the power of word-of-mouth, that is, people recommending the game to each other. At one point during the launch, everyone was talking about it. Even my 70 year old mom knew about it (and watched my 9-year old nephew playing it). You can imagine what such power does to your business both in good and bad.

And speaking of which, Telstra Australia has discovered from their data that promoters (people scoring 9-10 on NPS) deliver three times the revenue value of detractors (customers scoring 0-6 on NPS). Once Telstra understood this point, they successfully created use cases that drive impressive results in mobilising their promoters.

As a result, they have gained 10% market share and 90% share price increase over last 5 years, making them one of the most valued brands in Australia. So, who says all telecom services and companies are commoditised? Only those who miss the power of CEM. To avoid claims that they got lucky, Capgemini Consulting Analysis found these differences between telecom service providers who have a low or high NPS:

	LOW-NPS TELCOS	HIGH-NPS TELCOS
Growth over 2 years period	-7%	+33%
Competitive position	Mostly incumbents	Mostly smaller operators
Average age of the operator	21 years	13 years
Operating model	Mostly physical	Digital only and hybrid
Sales and customer support	Stores, call centres and limited digital channels	Primary digital, through mobile apps, web and social media

Based on a study by Ovum, there are several critical areas for telecommunication companies to focus on to improve their Net Promoter Score radically[106]:

- **Commit every department of the company to increasing NPS**
- **Improve customer service on social media channels**
- **Take more customer-centric approach to service contracts and warranties (the good old T&Cs)**

---

[106] Source: Improving Customer Experience Through Actionable NPS Strategies by Adaora Okeleke

- **Encourage a positive change for all employees, with exceptional performance recognised and rewarded, and the objectives of underperforming employees realigned to make customer experiences better**

With a well-designed Voice of Customer programme, you should know exactly what areas need improvement in your own organization particularly. And so we can continue to example use cases that you can consider as part of driving Net Promoter Score with your client base.

Refer-a-Friend

Enhancing your digital channels to encourage customers to refer one another seems sensible, yet few service providers do it. The first step in driving Net Promoter Score up is to make referring friends as comfortable as possible. Help people to invite their friends and give them something for their trouble, using referral codes, gifting and other tools - and they will do it.

A Personal Touch

A great way to encourage customers to recommend your services and products to their friends is to give them a personal way to do it, so enable them to buy and gift innovative micro-offerings for themselves and others. Create innovative opportunities for customers to gift each other and see your revenues soar.

What Is Mine Is Yours

As Facebook, Instagram and other social services demonstrate, people love to share – so why not share telecommunication services? And why limit sharing to families? If you want people to recommend your company to friends and colleagues too, enable sharing between any peers in your network.

Free Allowance for All

Operators in developing markets have used freemium models to get subscribers using mobile data. Having acquired a taste for accessing their favourite services, they often progress to paid subscriptions. An excellent example of this was at Airtel India, a company who introduced a promotion for Airtel Pre-Pay users for two months with up to 30 MB of mobile data dedicated to Facebook access.

Unlimited Video and Music Streaming

T-Mobile USA has turned itself into a self-styled 'uncarrier', rolling out unconventional but well-targeted propositions that customers love. Two examples of new T-Mobile propositions are 'Music Freedom' and 'Binge On', zero rating music streaming and video services on its network. P3 Group analysed the Binge On proposition and found it to be a win-win-win for customers, video providers and T-Mobile.

Customers spend up to 50% longer on video apps, so they get more from their mobile service. Also, video providers benefit from increased usage of their video services, and T-Mobile has grown its subscriber base from around 38m in Q2/2012 to 62m in Q2/2015 while keeping ARPU levels flat (cleverly, T-Mobile only offers zero-rated music and video streaming propositions to their middle and higher-tier customers).

As the video rate quality for the Binge On services is optimised to deliver just about 60% of standard quality, T-Mobile has even managed to carry more video traffic while maintaining or also decreasing the load on its network.

Make Customers Your Heroes

Happy customers are your best advocates, so why not give them a social platform from which to shout about their great experiences? A McKinsey study found that "consumer-to-consumer word of mouth generates more than twice the sales of paid advertising" and showed that clients acquired this way had a 37% higher retention rate. User-generated content is a trusted and authentic way to communicate the attributes of your brand, and you can build full-blown marketing campaigns and promotions around it.

It also provides a way for you to recognise and thank your customers as your marketing heroes. Influencer marketing also helps your search engine optimisation ranking. According to The Social Media Revolution, user-generated social posts account for 25 percent of search results for the world's top 20.

Mobile TV for All – Even Non-Customers

Creating a relationship with a potential client on another mobile operator's network can be very difficult, but that is the primary goal of Net Promoter Score. Most operators throw money at conventional advertising that emphasises price discounts and service superiority, but much of it falls on deaf ears. Swisscom took an innovative approach when they launched an open mobile TV product. Swisscom customers had this usage zero rated on their plans. Non-Swisscom customers were able to download the app by signing up via social media.

The Facebook sign-up allowed Swisscom to market to these customers and tempt them away from their existing operator over time. It also gave existing Swisscom customers a good reason to recommend their service provider to others. Despite fierce competition with cable service providers, the number of Swisscom TV access lines rose year-on-year by 14.2% to 1.33 million, over 60% of which use the cloud-based service Swisscom TV.

What a Pleasant Experience This Has Been

"It's not only what you do, but also how you do it."
- Might as well have been you or me?

Quality of Experience (QoE) looks at a service provider's offering from the standpoint of the customer. It asks, "*What mix of goods, services, and support would provide you with the experience you desire and expect?*" It then asks, "*Is this what the service provider has provided?*" and if not, "*What changes need to be made to enrich your total experience?*" Thus, Quality of Experience assesses human expectations, feelings, perceptions, cognition and satisfaction with products, services and applications. AppOptix (2016) conducted a user survey amongst Android mobile users, asking them about how different vital factors are when choosing a mobile service provider. The findings from that study were exciting as this is the percentage of respondents who found the following matters relevant:

- **85% - Quality of network**
- **70% - Quality of customer service**
- **18% - Used or recommended by friends or family**

Thus, the Quality of Experience (including network and customer service) is the top 1 reason for someone to choose a network, at least based on this one study. Several other studies have come to this conclusion also, typically having the Quality of Experience and customer service in top 3 factors along with price-related factors. So, there is no denying that price is not essential for the customers, but there is also a need for acknowledging that telecom companies need to service their clients well, too.

Were you surprised about the low importance of NPS when choosing a mobile service provider? I was a little bit. But let's remember that the research question here was what factors influence the decisions most, and it seems that people rate the customer experience and price above NPS. The previous does not measure how many more people have taken action as based on a recommendation from a friend.

When I have been conducting that study for various companies, that number is typically around 20-40% (in some businesses it can be much higher). Thus, while NPS is not necessarily considered as an essential factor when choosing a mobile operator, it does have a positive impact on influencing those customers to take action. Let's take a look at potential use cases that can affect the Quality of Experience for customers.

Speed Tiers

Customers who run out of data and are cut off will are very likely to report a bad experience. Operators with a remarkable network quality advantage over their competitors have an opportunity to press home their advantage with unlimited data combined with 'speed tiers' (variable MB/s). This model is common in Nordic countries - Elisa Finland is one prominent example. Swisscom also launched speed tiers - voice, text and data are unlimited, but tiers are differentiated and priced by speed. After an initial period of 'self-right grading', Swisscom speed tier customers returned a significantly higher ARPU (€6.6 per month) than customers on legacy plans.

Put a Cap on the 'Warez'

Operators limit P2P traffic and file-sharing for one of two reasons - because it's mandated by the courts (as in Ireland) or to minimise the impact of 'bandwidth hogs' and illegal software (warez) sharers. Being able to monitor traffic on both fixed and mobile networks is important as it gives operators the chance to throttle heavy traffic during busy periods, but allow it at full speed during off-peak periods. This keeps most of the operator's base happy as they have consistent speeds at peak times and encourages heavy users to download and share content during off-peak times.

Video Optimization

A third party video optimisation node can allow the operator to impose dynamic optimisation techniques based on criteria such as price plan, time of day, device or location. This can reduce the impact of video content on the network by 40% to 50%. T-Mobile in the

U.S. introduced Binge-On, which zero rated access to a wide range of video services. The stipulation attached to this proposition was that the available bandwidth for these services decreased to between 60% and 75% of pre-Binge-On levels – free access in exchange for slightly poorer quality.

Battling Network Congestion

As 3GPP standards evolve so too does the technology to control radio access network (RAN) congestion. To enhance overall mobile network performance, the policy control system needs to detect cell congestion dynamically, so that PCRFs can interwork with network probes to anticipate service degradation. Business logic can then be implemented to provide intelligent subscriber and network-aware analysis to trigger contextual information to the PCRF for subscriber control decisions. This latest advancement in the 3GPP space is called RCAF or RAN Congestion Awareness Function and provides operators with a standards-compliant tool to address congestion.

Network Insight

The most prominent pay-per-view (PPV) event of 2015 was the Mayweather-Pacquiao boxing match. Unprecedented demand meant that many cable operators struggled to provide customers in time for a fight which they had paid handsomely to see. Keeping customers informed of this kind of scenario is crucial to minimise social media backlash. Real-time automated communications keep customers informed – and off overloaded call centres.

Policy for Quality of Service

Different services need variable bandwidth to deliver the required quality of experience. When providing the best quality of service the operator needs to consider what the customers' expectations are against the operator's likely cost/return for delivering the traffic. For videos such as those posted on YouTube, for example, a customer may not expect crystal clear HD quality so the operator can reduce the speed of delivery, knowing that it will not affect the customer's expected experience. Alternatively, a higher QoS can be assigned to IPTV services, and this can be further differentiated depending on the user device.

Heavy User Offload

Tying network-switching technology (such as ANDSF) into a network congestion system gives the operator more control of resources related to their mobile data network. For instance, very heavy users who put an undue load on the network can be offloaded more readily to Wi-Fi, should an available hotspot be in range. This will protect the network and ensure a better QoE for other customers. Why does this matter?

The Copenhagen-based neuro-science company Neurons Inc. found one reason - stress levels in the brain rise from 19% to 34% while waiting for video buffering and that the cognitive load is similar to watching a horror movie! A stressed out customer is far more likely to churn. Service providers should also look into time and location-based offloading.

Operators that have specific congestion issues at certain times of the day can set up policies to more aggressively offload their mobile clients to Wi-Fi at these times. Location can also be used to control offloading. For example, a shopping mall in a high population city is prone to cellular congestion. More aggressively offloading customers to available Wi-Fi in this particular location will also boost the customer QoE.

Squeeze the Last Drop Out Of the OPEX

Many telecommunications service providers have grown cumbersome over the years and now find themselves struggling to remain agile while more nimble competitors pick off their customers. While complexity has built up in the products and services that telecoms offer, their customers look for simplicity of service across mobile, online and retail platforms.

In the meantime, the industry has to deal with the multiplier effect of each change in markets, products & services and technological infrastructure. As we already discussed earlier in this book, there is plenty of OPEX to be saved through Customer Experience Management in telecommunications. Many of those opportunities are better to find from your business, but in the meanwhile here are example use cases on impacting the operational expenditure.

Instant Customisation

Letting customers customise their own plans means that instead of switching to a competitor or calling the call centre, they can add or change whatever they need through an intuitive user interface. For example, if a customer is watching videos and has run out of their data, let them top themselves up easily and quickly.

Control Sharing

Keeping tabs on who is using how much of shared resources can be a tedious task for the customer. By setting limits and notifications for each user in the shared pool, the customer can efficiently ensure that one misbehaving member does not use up all the milk in the data carton.

Time-Based Plans

Simplifying services is a crucial part of reducing complexity. For example, what's easier to understand - a minute or a megabyte? According to customers, many customers fail to understand the meaning and value of a megabyte or a gigabyte. A more straightforward approach adopted by Bell Canada offers five hours viewing of their Mobile TV proposition for $8 per month (with additional viewing available for $3/hour) – a proposition that subscribers can easily understand. A similar implementation of this approach would be to offer 1 hour, 8 hours, or 24 hours of data access without metering.

Direct Operator Billing

Enables digital content to be easily invoiced via a one-click-payment that is added to the customer's next invoice or deducted from their prepaid credit. The operator receives a portion of the charged fee, helping to grow their digital services-driven revenue. This model has advanced on significantly from the ringtone-billing days, and an excellent example can be found in Malaysia, where Maxis has launched operator billing for Google Play app purchases.

Customer Experience Based Charging

---

*"Quality is remembered long after
the price is forgotten."
- Gucci Family Slogan*[107]

Customer experience based charging (CEBC) is a use case of its own that is mainly addressing churn and customer satisfaction. There are different ways to implement it, both from business and customer perspectives. Let's start with the business perspective and then move on to the customer one.

Pricing is a complicated topic in telecommunications. Companies work hard on their tariffs and pricing models. And many times it seems to be also the weapon of choice when competing. Unlimited bundles are one example where there is not much more left on the feature side to contend with than price.

CEBC takes an opposite approach. When done from the business perspective, the service provider will identify those areas that are valuable for the customers (and for which they are willing to pay for) and take those out from the bundles and charge them separately. Maybe an example of such a service could be that a customer pays extra for not only having access to Netflix service but also gets the best available bandwidth for it. This way the customer would get added value on top of the access to service and likely wouldn't mind paying extra for it.

From the customer's perspective, CEBC could be a way for a service provider to put their money where their mouth is. Perhaps an operator who boasts about the best *"4G coverage in the country"*, but that is not the case in some areas of the country, the customer will be

---

[107] Source:
http://www.searchquotes.com/quotation/Quality_is_remembered_long_after_the_price_is_forgotten/237862/

charged less (or given some other form of remediation). It would work almost like a *"money-back guarantee"*. Such an approach connects B/OSS insights with solving customer issues close to the network source where a problem may have occurred. It could enable close to real-time solutions to problems.

For example, by monitoring customer experiences in real-time, it is possible to find interactions that are unacceptable or below pre-defined quality KPIs. Then it is possible to take mitigating action immediately for the customer, such as:

- **Refunding part of the service cost (e.g. $10 refund)**
- **Adapting a rate plan for a period of time (e.g. 50% discount for a week)**
- **Offering a top-up (e.g. 2GB data)**
- **Proactively notifying subscribers of a problem or remedial action they could take (e.g. reboot your phone/router)**

Assuming that you have the technology in place to monitor the QoE in real-time and take action based on it, the next challenging question is how those quality KPIs should be set. That will depend on your customer and business strategies, network performance and many other factors. You could use for example:

- **Real-time threshold event notifications (e.g. data speed drops to 2G in a 4G area)**
- **Information sourced from Voice of Customer and business analytics (e.g. prediction)**
- **Information sourced via Deep Packet Inspection (e.g. virus detection)**
- **Feedback from partners (e.g. Netflix saying that your network is performing less than the competition)**

Customer experience based charging is not for the faint-hearted. It is an excellent option for those service providers who want to be leaders in their market through customer-centricity and who are willing to take the pain to learn what will work for both, customers and the business. For others, easier use cases may turn out to be a more comfortable start.

Get Rid of Those Bloody SIM Cards Already!

This use case is a simple one from the consumer perspective, much less so from the communication service provider's. To be honest, we should ask ourselves the question whether swapping SIM cards is actually any problem for the consumers. At the end of the day, people don't change their SIM cards that often, do they? And that is true from the physical perspective of this matter. It is not hard to wait for a couple of days before receiving the new SIM card in the mail.

A bit bigger pain point is finding that small pin that is needed with most smartphones today. At least I manage to lose it immediately after I have popped in the first SIM card into a new phone. However, the more significant change here is the mindset. The old-fashioned SIM card technology helps operators to make the customer experience of switching operators harder. If they could do that quickly in their phone, surely that would drive churn higher?

For those operators who have customer experience issues, it most certainly could do that. For those who offer a great experience, it might be less of a problem. It's not like an average consumer spends all day long looking at opportunities to switch service providers, they only do that when they feel like it. So, whether you are a winner or a loser in this matter is up to you.

Putting the mobile phones and tablets aside, other new technologies will increase the need for digital SIM cards. Various consumer smart devices like watches, home security, and appliances are asking for network connection nowadays. Many of them still connect to Wi-Fi, but an increasing amount would benefit from a mobile network connection. Apple smartwatches have been leading this development in the recent years.

It is not beneficial to go into the technical details of how all this would work as the standards have not been fully agreed yet. This will be a big technological and political push for service providers. And yet, as the popularity of Internet of Things (IoT) grows and the choice of connected consumer devices becomes more significant, it just has to be done.

Please, Don't Leave Me!

"Every company's greatest assets are its customers because without customers there is no company."
- Michael LeBoeuf, an American Author

90% of the telecom industry measures customer loyalty through churn while 89% of consumers are willing to switch providers as a result of a bad customer experience (Source: Forrester Research). Also, Ovum conducted a Global Consumer Insights Survey on reasons why customers churn with mobile service providers and 25% of those consumers said that they are leaving the current provider to get better customer service from someone else.

Even the slightest reduction in monthly churn creates enormous financial benefits for the operator and an increase in the lifetime value of their customers as we already discussed earlier in this book. For example, the average value of an AT&T and Verizon customer in the United States is around $3,000 over a four year lifetime. With over 100 million customers in their base, that's a great deal of money.

Yet an average of 67% of telecom customers churn because of disappointing experiences with the operator (Source: Esteban Kolsby). We need to acknowledge and address the issue of bad customer experiences and low emotional connection with operators, and its contribution to higher customer churn. The benefits of doing that are evident: wireless carriers in the United States alone could boost their margins by 1% of revenue through 2 percentage point improvement in churn[108]. Here are ideas for use cases that could impact churn for your company.

---

[108] Source: Strategy Analytics' Service Provider Strategies, 2017

Apologise For Poor Coverage

Analysys Mason's connected consumer survey in 2015 listed poor mobile coverage as the second highest reason for churn behind poor customer service. Tracking dropped calls and responding quickly to poor network experience is an efficient way to mitigate customer anger. For example, if a client is seen to have dropped five calls in a one-hour window, the service provider should make an immediate offer – perhaps of free airtime minutes.

This is particularly important in lower ARPU markets where relatively small bundles of voice minutes are still common, and carefully husbanded. This kind of experience-triggered offer will enhance customer perception of the operator, even when coverage is less than stellar.

Broadband Outage = Free Mobile Data

Over recent years, broadband has become as important to consumers as water and power utilities. If power is cut during a storm, customers generally accept that such things are inevitable. When broadband goes down, the storm that ensues is all over social media. A multi-play operator can utilise their entire infrastructure to mollify customers.

For example, a fixed broadband outage could trigger a notification to affected mobile subscribers that their mobile data access will be zero-rated for 24 hours? Or a notice to customers advising them how to use their mobile as a portable hotspot in the home - not only allaying customer anger but rewarding customers for having multiple services with the one operator and encouraging alternate use for the mobile device?

Recurring Flex Plans

Giving customers a fixed amount of data can limit usage. Bell Canada has an innovative structure for their tablet data plans. The bundle price charged at the end of each month reflects the actual amount of data consumed in that month, not just the initial bundle price. This has the dual benefit of giving customers flexibility in how much data they can consume while at the same time helping them always get the best bundle price.

Zero-Rated Content

Multi-play providers with access to TV content have successfully offered TV to their mobile base. A1 in Austria charges clients €4.90 a month for access to a mobile TV. The usage is zero-rated and doesn't reduce their monthly data allowance. This type of zero-rating works very well with customers and gives them another reason to stick with that operator longer. Similarly, T-Mobile in the US has had great success with zero-rated offerings such as Music Freedom and Binge On.

The Rolling Data

'Rollover Data' typically allows a customer to roll unused data allowance on to the next month. As part of its rollover offer launch in Australia, Virgin Mobile conducted research into 1,000 consumers and found that 94% of Australians thought it unfair that something they had paid for was taken away with no chance to use it. Not surprisingly, they loved Virgin's proposition.

T-Mobile US reduced post-paid churn from 1.73% in Q4 2014 per month to 1.3% in Q1/2015, attributed largely to allowing customers to roll over their data. If you take the winning rollover consumer proposition and turn it upside down, does it still make sense? In the case of Roll-under Data, it does. Not only can you roll unused data over into next month you can also borrow data from next month if you use up all your limit. This was launched in South Korea by KT Telecom, branded 'Data Push and Pull' and has been a runaway success.

Up-sell and Cross-Sell

Multi-play operators know that the easiest way to drive down customer churn is to sell them more services. The more services a consumer has, the less likely they are to leave – and if the offers are well-targeted, the customer's experience is improved too. A central European operator released churn rates which illustrate the drivers for operators: four product churn was 3% per year versus 21% for single play mobile (image below).

Single service	2 bundles services	3 bundled services	4 bundled services
1	2	3	4
CHURN RATE OF	CHURN RATE OF	CHURN RATE OF	CHURN RATE OF
21%	13%	8%	3%

WE CAN LEARN FROM OTHERS, TOO!

In 2017, Openet[109], together with European Communications, surveyed attitudes to customer experience in the telecoms industry, which gave pretty impressive results, prime among them: customer experience is a vitally important success factor in the competitive telecoms market and communication service providers are quite bad at customer service.

As we have already discussed, Customer Experience Management is a paramount success factor in the highly competitive telecommunications industry. At a time when clients struggle to differentiate service providers by price, service offering or network coverage, nothing is as effective as the high Quality of Experience.

The reasons for telco companies having challenges in responding to this customer need are naturally manifold. Sometimes these reasons (historically at least) are entirely rational because, as an industry, our focus has been in engineering and technology. Our attention is taken by the physical - the network, the handsets and all the other gadgets that are required to keep a telco working. Traditionally, customer service made very little difference to telco success.

For almost a century, most fixed operators were monopolies or cosily co-existing duopolies. You don't need an MBA from Harvard (or even a PhD from Turku School of Economics) to know that monopoly power and customer experience are uneasy bedfellows - customer service typically comes down to a simple *'here it is, take it or leave it'*. Even the launch of mobile communications (which most governments were determined to ensure was an open and competitive industry) made a small impact - demand outstripped supply for so long that almost all CSPs could turn impressive profits on the new technology without having to try too hard to 'delight' the customer. Unfortunately, that is still the case in many markets.

Also, telcos are kinda mature businesses. The old mobile operators have been successful and very large-scale consumer businesses for a quarter of a century. During that period they have

---

[109] The basis of this chapter was originally written by Bob Machin, a marketing consultant at Openet. It was also published in an Openet whitepaper and at Disruptive Views website.

also been very successful in building up an entangled spaghetti of information technology (itself evolving fast over the same time) to support their customers. Multiple systems, of varying ages, mainly designed for workhorse performance rather than real-time responsiveness, do not respond well to demands for agility.

And migrating all those back-office solutions to something more livelily is not something that most telecom companies contemplate lightly. None of this carries well with customers, of course – they just want something that feels good and works well. Preferably an experience that gives them what businesses such as banking, retail and travel seem capable of providing in the 21st century.

Let's continue with having a look at other industries and businesses with a reputation for delivering an excellent customer experience. For argument's sake, I will define excellent CEX as *'meeting or exceeding customer expectations in a way that provides competitive differentiation in the telecommunication market'*. We will see that outstanding customer experience isn't always about high-tech (though a little doesn't hurt).

Sometimes it is about designing and implementing the business cases more imaginatively to lure and then retain the customer. So, here are six customer experience strategies that seem to work outside of telco sector, and thoughts on how they can be applied to telecoms – particularly as telcos diversify into ever-broader service offerings.

Tiered loyalty programmes

Similar to that offered by British Airways, hotel chains, and other airlines. Basically, more usage means more points. Points get you flights and upgrades. Note that airline frequent flyer programmes don't necessarily equate to love for the carrier, but they are remarkably effective in keeping the customer's business. And customers like them – the average American household is enrolled in 18 loyalty programmes (though only active in eight).

Such a simple points approach could work for telcos, too. Start off just offering points for spending. Then vary points to foster the use of new, or different products so that the loyalty scheme became a way to impact client behaviour. Provide Bronze, Silver, Gold, and Platinum-level (maybe even a secret Black-level?) memberships to reflect customer longevity and spend. Allow customers to trade points for service payments, phone and upgrades, and special promotions. Your points-based loyalty scheme could be introduced as an adjunct solution, driven mostly out of charging and billing.

Loyalty cards

For example, by offering every 10th purchase free – as provided by Starbucks, other coffee shops, hairdressers, and other firms with which customers tend to have regular engagement. Every tenth coffee, haircut, whatever, is free. A telco could do something along the same lines, but without requiring a physical card – for example, offering customers every 4th GB of data, or every 12th month (ideally the renewal month) for free.

Customers could then use their free data automatically, or rack it up to use at their discretion. Such a virtual loyalty card system could be introduced as an adjunct solution, driven mainly out of policy control and charging and represented by a mobile app and web solutions.

Appeal to quality and good taste

High-quality approach sustains famous companies like John Lewis, Fortnum & Mason, Bloomingdales and similar middle-to-high-end retailers in many countries. And 88% of respondents to the recent Clickfox survey[110] indicated that quality is a critical factor in their decision to remain with a brand – even more than customer service. The key? Unostentatious, unshowy quality that is appealing to clients who are not price-driven.

Telcos would look at their range of offerings, perhaps limiting it to a small range of excellent, non-flashy handsets. Service bundles would be pleasantly effortless – not the cheapest, but straightforward. The focus would be on service quality, competence, the maturity of approach. The customer would always be right. Such strategy might be difficult to achieve in the short-term for traditional communication service providers while being more comfortable for a new operator or Mobile Virtual Network Operator.

Develop a closed ecosystem

Driven by tech and unpretentious style, such as that which sustains Apple and makes devotees of their customers. The offering is 'inwardly compatible', outwardly not. Customers become voluntary captives in exchange for something amazing. It may be hard though to imagine a telco (or any other business) coming up with the

---

[110] http://www.loyalty.de/wp-content/uploads/cf-survey-results-brand-loyalty.pdf

combination of technology, design and 'X-factor' that has created Apple's extraordinary customer loyalty. Therefore, other options might include exclusive Apple reseller relationships, hosting an Apple MVNO venture, or being acquired by Apple. Maybe this is just a daydream, but believe me, the day will come when this will be done by someone.

Membership

Football clubs like Schalke04 and private members clubs in London and New York establish a psychological tie-in which becomes challenging for members to break – even when (for football fans at least) the customer experience is less than amazing. In this strategy, telcos would look for natural affinity groups and aim to engender a sense of belonging, mutual support between clients and identification with the brand.

It would almost certainly require telcos to establish sub-brands and target younger consumers (pre-customers) who would be susceptible to subtle offers to develop the coolness of the brand and tribalism. Implementation of such approach could be low cost, but also a long game, which would require very nuanced marketing and dedication (think GiffGaff in the UK). 'Membership' systems could be adjunct to the existing IT and relatively low-tech.

An unashamedly high-end pitch

Of the kind that, say, Burberry or Gucci offer, to people who have significant disposable incomes and like to advertise the fact. This has been done before, at least to an extent, with Nokia's Vertu, but the market could perhaps sustain another better-implemented offer, mainly as handset technology and evolution has somewhat stabilised.

To implement this strategy, create a sub-brand for extremely high-end matching equipment, personal concierge services and so on. This would require high capital investment in stock, personalisation, personal services and well-focused marketing. So, you might want to be serious about it if you start it.

We have discussed six strategies, some of which are more realistic today than others. Highly performing Customer Experience Management is a challenge for the telecoms industry, but part of the difficulty may be that the telecoms sector is too introspective. Perhaps, instead of looking for better, more engaging ways to do the things that telecoms already do (which rarely seems to impress the customer hugely), it should be looking for very different things to do.

The example set of strategies and use cases that we've collated here work in adjacent industries which face many of the similar challenges as telecommunications – competing in a limited market and endeavouring to maximise the lifetime value of the client. Could any of them work for your business?

HEAR IT FIRST-HAND: TALKTALK BUSINESS

"We zig when the world zags!"
- Danny Sullivan, Director, TalkTalk Business

Most telecommunication companies are working hard on customer experiences to gain a competitive advantage, and typically we think about consumer-oriented, B2C, customers when we talk about it. However, Customer Experience Management is as essential on the enterprise side, too. Danny Sullivan, Director of Enterprise Service Management at TalkTalk Business (TTB), was kind enough to share with us his views on this.

Danny works for the COO office and delivers services to TTB's top 147 B2B accounts in the corporate, wholesale and system integrators markets. He manages very successfully a service management team of over 20 Service Managers driving up revenue, customer satisfaction and supporting sales with new large customer engagements. He has been one of the key contributors to TTB's service model and implementing impressive results (from 56% to 72% in less than two years) in increasing customer satisfaction (measured with a bespoke SmartSAT metric).

TalkTalk Business serves over 180,000 corporate customers and works with over 600 partners across the United Kingdom. It is part of the TalkTalk Group that has revenue over £1.8 billion and is one of the FTSE 250 companies in the UK. From a broader perspective, TTB's journey to leading the B2B market in the UK started with a decision to become a market disruptor in 2006 and it transformed into a customer-focused company around 2014. Today, TTB is a market player who offers its customers excellent value for money with a network that keeps customers happy.

The business TTB is in is at risk of becoming highly commoditised as the business model is a relatively simple proposition of enabling connectivity for enterprises. This means their business

176

can be done in high-volume, high-value manner. Such an approach requires technological sophistication that can be scaled. The network needs to ensure high Quality of Experience for the customers while managing costs and effortless experience with automation.

Though the service may be simple, customer needs are not. Some of their customers typically want to manage the network themselves, leaving reasonably little for TTB to do at the user's end. Some of them need more managed service, including a router in the office. Even though the product is the same for both customer segments, the customer experience is very different.

That creates a high-level of complexity for TTB to deal with. They have to differentiate the service for each customer segment, but at the same time, avoid developing different systems and processes for all segments. Such needs call for an unprecedented business and technology agility. It is a constant battle of simplicity and complexity.

The B2B sector is not left without the same challenges as the B2C sector sees when it comes to customer expectations. Most telecommunication companies have CAPEX, OPEX and Product Development cycles that are way too long compared to what customers want. TTB has set as its goal to be able to launch new services in just a few weeks (or a couple of months) cycle, instead of 9-12 months that it typically takes for telcos.

That is very challenging to deliver with legacy technology. The expectations from customers and partners are to benefit from new technologies at the lowest prices possible as quickly as they can. Such desires leave the telecommunication companies to juggle with creating new services, dealing with legacy services while keeping the customer experience great through the changes.

While dealing with customer expectations is similar between B2B and B2C sectors, there is a big difference in understanding and retaining customers that have the highest value. While in B2C sector customer numbers are typically in dozens or hundreds of millions, on the B2B side Average Revenue per Account (ARPA) is significantly higher. Thus, customer experience increases even more in its importance as losing even one key account can be devastating for business results. Danny elaborates this further:

> *"B2B customer relationships are much more delicate as the account numbers are small. Sampling and measuring of customer experiences through surveys can be misleading or take the focus into wrong areas. You have to check the relationship status with customers on an individual basis, not just with surveys, as they can miss important feedback, which you only get through being intimate with customers."*

And that is why Danny says that even in the B2B sector, *"it is people, not the systems"* that matter the most. The people are the ones who keep the wheels on the bus rolling when things get hectic. Today's companies are juggling with several projects, many teams, and technology changes that both aid and hinder the work. This people-focus has worked well for TTB.

They have progressed from being one of the lowest in the industry to being above British Telecom and are on a trajectory to surpass all other competitors. Perhaps that is also why TTB's customers are most satisfied with their Account & Service Management, Products, Portals, and focus on the business market. As mentioned, the results are impressive as the overall trend for the TTB's SmartSAT customer satisfaction metric looks like this:

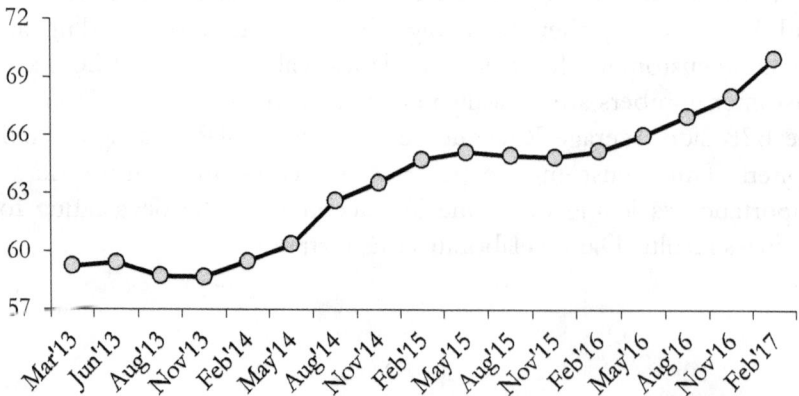

Any company boasting four consecutive years of improvement in customer satisfaction trends can give themselves a big pat on the back. One should never take it for granted as customer satisfaction is a fluid thing, moving as the world, expectations, technology, etc. evolve. Danny's advice for us on how to do it is to focus on what customers value, not what we think they might value.

He also warns of us about not getting the "analysis paralysis". Instead of overanalysing everything, TTB focuses on quick launches that enable testing and learning together with customers. It is enough to understand only 80%, no need to kill yourself by trying to be always 100% right as it will change soon anyway.

"MARRIAGE" ADVICE BY ENRI-K SALAZAR, EXECUTIVE DIRECTOR, OPENET

Sometimes relationships come to an end. In the telco world, we call that *'churn'*. Too many times it is just a number on a balance sheet. For many it is something to get rid of, a thing that hurts our finances. However, we should not forget about the strong and unpleasant feelings that come with leaving someone. In this chapter, Executive Director Enri-K Salazar from Openet goes through some of the most common feelings customers experience with their Digital Service Providers, when churn happens.

It's Not You, It's Me

Surely you know the old cliché breakup excuse: *"It's not you, it's me!"* However, in the case of a customer leaving their Digital Service Provider, the answer would more likely be: *"It's not me, it's you!"* That is because the customers are the ones calling the shots and many operators are failing to serve them. They have needs and wants that have to be met, customers' lifestyles and circumstances change, and they are in need of constant adapting to those. Customers expect to be treated with respect, you to know and understand their needs, to send them personalised and contextual communications, to enable them to react, and to manage any add-ons whenever they want. Since this is the case, it's you, dear Digital Service Provider, that I am breaking up with for not considering me well enough as your customer.

It's Just Not the Right Time for Us

Right timing is vital. Once a customer has contacted you to get help, and if the help is not given promptly, that is a perfect reason for the customer to breakup with you saying *"It's just not the right time for us"*.

"82% customers say that getting their issue resolved quickly is the number one factor to a great customer experience."
- Erin Kang, LivePerson

Further on, when your customers receive endless texts, emails, and notifications with irrelevant offers and other communications that have no value to them, don't be surprised if they again say *"It's just not the right time for us"*.

The Breakup Text

You may have heard of a breakup letter, email or even an SMS, and if you were unlucky enough, you may have received one yourself. Either way, it is never well received, the sour feeling it leaves still aches long time after. Well, we could say the same thing happens when a Service Provider tells you that your fair complaint is not accepted or compensated. Whatever the reason, negative communications from their Customer Service department are never easy to receive, especially the ones that start with a *"We regret to inform you…"* heading. Even before reading the rest of the communication, you know it's a *'NO'* answer, which creates a negative feeling before even getting into the reasons. Communicating an unfavourable outcome to a customer is never an easy task, so Digital Service Providers need to think twice about the methods and channels used to do this. A negative communication is almost always a bad experience. Is it fair? No, but unfortunately, this is what customers feel. As in any relationship, talking about the situation will at least give clarity and a sense of caring. It can make the negative emotion less hurtful. So, instead of just sending the negative letters, text or emails, make a call and explain the reasons for why the complaint is not valid. You can even turn the negative communication into a positive experience, by offering something valuable as a token of appreciation, showing you care and that the customer matters. After all, 37% of customers are satisfied with service recovery when they are offered something of monetary value (e.g., a refund or credit). But when the business adds an apology on top of the compensation, satisfaction doubles to 74%[111].

---

[111] Source: https://www.groovehq.com/support/customer-support-statistics

Marriage Of Convenience

By definition, '*a marriage of convenience*' is a marriage contracted for reasons other than that of relationship, family, or love. Instead, such a marriage is orchestrated for personal gain or some different sort of strategic purpose. How many customers feel they have a marriage of convenience with their Service Provider? I'll give you an example. I have my mobile service with Eir, because I get a discount through my employer. In July 2016, I received an invoice of $10,400 from Eir. Despite many of you know what a Bill Shock is, it appears my mobile operator didn't. When calling to find out about the extensive charge, the answer was "*Let me pass you to our Credits department for further discussion*". After explaining to them what Bill Shock means and that I was not paying for the full charge, as they should have blocked me before accumulating such a cost, their answer was still "*OK, but let me pass you to our Credits department, they will be able to help you.*" After complaining to our company's Account Manager, he asked me to stop the direct debit from happening and he "*would deal with getting it sorted*". That started a year long, tedious process. After calling him every week, writing two emails a week for status, and getting all sort of lame excuses, it finally got removed from my balance in July 2017. And yet, while writing this chapter at the end of 2017, I am still a customer of Eir. Why on earth? Because it is a marriage of convenience as mentioned earlier. Would I wish it wasn't? Yes, of course, I wish they would do a bit more to treat me like a human being. I wish they would consider my relationship as something valuable and not a strategic one based on a company plan, but that is not the case. So I ask again, how many of your customers have a marriage of convenience?

The Silent Treatment

The worst of all marriage hiccups... the one that does the most damage, hurts most and leaves irreparable wounds... The silent treatment is just the easiest way to make anybody want to just disappear and forget the relationship ever existed. It will leave the sour taste that you just don't want to remember. And yet, it is the most common one with Digital Service Providers. It amazes me that the more I read about Customer Experience, the more frustrated I get with this simple fact: 67% of consumers want a response to their customer service questions within 24 hours[112]. Any customer that makes a query, a call, a cry for help, is because they need something. If that call or cry for help is not answered immediately, then why is it so surprising that the customer looks elsewhere for those answers? The silence treatment when a customer has a query can be a trigger for that customer to take the action to leave, the last straw that will bend the scale over to the leaving the decision. The silent treatment means the exact same to your customer as when it is used as a breakup strategy: "*I don't care about you, leave me alone, go away, and go find someone else to bother*" …and so they do, your customers will leave.

HAPPILY EVER AFTER

---

*"Customers don't expect you to be perfect. They do
expect you to fix things when they go wrong."
- Donald Porter, Vice President, British Airways
(the reader can decide how well they live up to this)*

There's an old saying: all good things must come to an end. However, for us, this is just the beginning, regardless of this book running thin on pages left. Customer Experience Management is not a project or programme, it is a process. It lives from continuous improvement and added value for the business and the customer. I thought the best way of ending this book would be a story that speaks to the theme of this book: marrying our customers.

There are plenty of analogies between CEM and marriages as we have already seen, and maybe that will help you to communicate the principles and their importance for your colleagues going forward. Let's get to our final story in this book...

It all starts when we are (young and) cheerful. Being single has been the best and worst times of our lives, as has been married, too. When you are single, you are the master of your own life. No need for compromises and taking others into consideration.

We can have fun and go out with our friends whenever we want. Oh, wouldn't some married people give anything to get that level of freedom again? And yet for some of us, it's also been very stressful. I don't believe people are created to be alone (though for a few it is the best option). While single we can focus on ourselves and our lifestyle, though it may be lonely, too.

Being single almost reminds us of having a start-up or a monopoly legacy telco business. Such businesses typically focus on doing something they find engaging with other people's money. Whether that sponsor money comes from generous venture investors or locked down customers is irrelevant. Where the focus lies is on the innovation and products of the business. That is generating organisation-centric value rather than customer-centric.

And it does work well for companies with limited market movement (perhaps in telcos it has worked for longer than in any other industry, except financial institutions). The challenge comes from a lack of genuine connection with customers and the market. Thus, marketing and customer acquisition costs tend to rise up, new competition arises, market share shrinks and fluctuates, and investor money runs dry.

This is the time when people start looking for a companion. Do you remember those times when you used to date? Or perhaps you are dating right now. It's quite exciting, isn't it? Also, a bit scary and at times awkward. You know, when you say something silly or get dumped with an SMS. Yet most of the times it is enjoyable dinners in good company, whether or not leading to romantic outcomes. The heat of the dating can be abundant. Big emotions, both positive and at times negative. Yet, the idea of being together with someone special is tantalising.

So it is in telco companies. Leaders come and go (e.g. see how many times Ericsson changed its senior leadership in 2017). The owners are looking for better and more from the business. The targets for growth and profit get bigger and more challenging to reach. Customer churn gets bigger as customers come and go even more than before.

The company keeps on creating customer expectations, leading to efficient campaigns, just to see later that those promises are hard to live by and thus the customer base starts to look like a Swizz cheese or a leaking bucket. But there are great moments, too, no doubt, it's just they don't typically last for long.

Time flies and the special one has been found. It's time to make commitments, get engaged and finally married. Move in together and agree who gets which cupboard. This is also typically the time when the realities described in Mark Gungor's video become a reality. Men and women think differently. And that's the beauty and the horror of it. In most cases, the finances of previously separate entities are starting to merge, and decisions are not made alone anymore, but together, typically in a form of compromise. The heat of the dating begins to fade away and turn into more sustainable caring.

In companies, at this point the board and senior leadership start to acknowledge what Peter Drucker already said more than 50 years ago, the customers indeed are the reason for a business to exist. Customer service is in the foreground of company's interests, and it is known that employee engagement leads to customer engagement.

The company has a clear understanding of how its economic situation is driven by customer satisfaction and advocacy.

Collecting bad revenue (e.g. penalty charges) is not anymore a significant revenue source. Also, marketing has changed from big value campaigns to adding relevant value for customers, in a way that they are happy to pay for. Instead of locking down customers with long contracts, they want to stay with the business as it is an excellent choice for them. Existing customers won't be penalised for getting new customers in, but all are treated fair and equal.

It is time to ask the second big question: how many kids will we have? There is a built-in urge in people to procreate. And so they will, creating a new reality. No more movie nights, dinners or hanging around with friends. Nappies, long nights, vomit and crying has come instead. How is it that the best and the worst thing in your life happened at once?

And were you prepared for it? You are lying if you say yes. It is one of those miracles that can't be learned from others but has to be experienced. But nothing on this earth lasts. Children grow up. It's time to learn to play football and do make-up. Time to unlearn the selfishness and to find the joy of sharing with and serving others. The pleasures in life now come from those small things.

Companies reaching this stage know (refer to the maturity models discussed at the beginning of the book, if you will) that customer-centricity is a lifestyle. It is not a transformation programme of a CEX project. It is a way of managing the business with the customer in the centre. Eventually, it may be time to expand the value proposition for customers with partners.

Mobile service providers, for example, have started to partner with OTT service providers, such as Netflix, Spotify and others. This makes a partnership where one offers the connectivity and the other content. Undoubtedly, some telco companies will look into the different type of alliances too, becoming content producers (or banks like Orange) themselves.

These help to reduce churn and create longer-lasting customer relationships by offering more than just a connection. The ownership of customer relationship becomes blurred, and it is understood that the customers are owned by themselves, not by any company, and therefore it is possible to function in a value network, without getting overly jealous.

Life's all good and jolly. Children take up a vast amount of time, and our own desires are put on the side. Everyday life becomes perhaps a bit too much casual. A yearning for freedom, self-expression and fun raise its head. It is time for a mid-life crisis. Everything goes from the local pub's luvvly-jubbly[113] *lady to yoga to buying a fast, convertible sports car (that's why I got one before a crisis, so it should come to me milder if I choose to have it). People try to find meaning in the void. So, divorce it is! Time to move on.*

Companies who see a significant uptrend in their customer KPIs (e.g. NPS, CSAT) may find it hard to sustain if it is not embedded in their culture. Money comes in, numbers look good for now. It's time to start to take it for granted. Investments and focus on Customer Experience Management begins to lose pace. Maybe it is time to reorganise and save resources from the grown customer operations department. But the world around keeps on evolving even though your company stagnates.

Customer expectations change, competition learns, and new technology creates possibilities. Customers start to see the grass greener on the other side of the fence and churn picks up again. Revenue, sales and customer acquisition begin to creep back as main discussion topics instead of customer value. Some partnerships start to fade away into the background. Ambitions for revenue and profit growth have increased over the years and begin to look somewhat unrealistic again. Thus, it is time to take back the short-term means for upping the sales.

Old is gone, and the new is in! Like a fresh breath of air, it is time to be young and wild again (well, at least to feel like it though the years may have had an impact to the body). Back to dating and looking for one's self. Some manage to find themselves, some even love. Some will keep on searching for the rest of their lives. Maybe it is time to get married again to ensure someone to have for the retirement days. Life may taste sweet again, but not as great as it maybe once was. You can't beat the best with good. Or take back what is done.

For companies, revenue focus takes over the customer-centricity. Ways that used to work for the customer don't anymore, and there's no real desire or budget to close the gap, so it keeps on widening. Partnerships get worse every year (like VitalyHealth's benefits and partners in the UK). Price wars become relevant again as customers see the less added value. The company still is good, but its glory days are gone. Interestingly, it doesn't stop some of the employees to still live in those days when things were better, the time has stopped for them (in their heads).

This doesn't have to be how the story ends, however. What if we took a

---

[113] Another British way of saying lovely. Made famous by the TV show 'Only Fools and Horses'.

different path when the everyday life starts to kick in? Instead of losing the passion, we would do all in our power to keep it and make it grow. Yes, against the current circumstances. We make sure that the family gets balanced time together as well as separately. Instead of children taking over the whole life we enable each other to have a personal life, too.

Instead of looking for answers from the outside and from the circumstances, we start to look for them from where we are, from the within. What would add the most value for all of us without forgetting the individual? We could live, learn and keep on growing together and skip the whole mid-life crisis and all the hardship that comes with it!

Instead of growing fond of our current ways of doing business, we could stay on our toes. We could choose not to take the customer for granted and to acknowledge that what we do today may work right now, but to keep on evaluating what would work even better. Also, we can prepare for tomorrow by focusing on: the as future is created today, not sometime later.

We can keep on working on adding more customer value and deepening our relationships with our partners, too. We can find even more valuable partnerships and instead of cutting, we can find ways to add without increasing costs. We become better, day by day, through growing our understanding of our customers. They are our friends, and as such we love them. And all this manifests itself as sustained growth, improved customer KPI trends, an ever increasing number of positive customer testimonials and a great vibe amongst the whole value network. How amazing is that?

And so you will be the master of your destiny.
Whether you will live happily forever after is up to
the choices you make right now.
How's it going to be?

ABOUT THE AUTHOR

Dr Janne Ohtonen has delivered various challenging Customer Experience Change Programs, several of which included double-digit performance enhancement. He holds a PhD in Business and Information Technology and has contributed to several scientific research projects. Dr Ohtonen's approaches are acknowledged as Thought Leadership and are used in hundreds of organisations around the world today.

Dr Ohtonen believes that the most impresive transformations to an organisation's profitability, efficiency and performance come through the alignment of Customer Experience with Business Processes, Enterprise Architecture, and Innovation. This requires leadership and commitment, both of which he provides expertise in. Dr Ohtonen is a regular speaker as a Keynote Speaker to C-level roundtables, MBA courses, and international conferences. He delivers practical knowledge in Customer Experience Management, Principles of Leadership, and Innovation and Change.

He has worked with well-known companies such as Apple, Avios, British Telecom, Pfizer, IAG Group, Satmetrix, DST Systems, Aviva, Metro Bank, Liberty Global, and British Airways. Dr Ohtonen has also published several books on customer-centric business process management and capabilities.

He typically engages with:
- Board members & C-level leaders looking to achieve a substantial improvement in Customer Experience and Loyalty.
- Top executives who need Voice of Customer programme and Customer Advocacy expertise.
- Business leaders who want to maximise results and overcome common Customer Experience challenges.
- Business owners who need an expert to deliver dramatic transformation in organisation's efficiency, profitability & performance through world-class Customer Experience.

Dr Ohtonen helps top business leaders to:

- Lead turnkey customer-centric change initiatives from concept to practical delivery.
- Support the building of a customer-centric innovation culture.
- Design and implement high-performing Voice of Customer programs.
- Create strategies to improve business efficiency, profitability and performance.

Dr Ohtonen was born and raised in Finland. He now lives with his wife and son in London, United Kingdom, known for its international business ambience. He is a Certified Scuba Diving Instructor and enjoys exploring the underwater world. A portion of his pre-tax income is donated to charities creating a livelihood for the poor.

CONNECT WITH THE AUTHOR

LinkedIn:
http://linkedin.com/in/janneohtonen

Twitter:
https://twitter.com/ohtonen

Facebook:
http://www.facebook.com/successconsultant

Email:
janne@threecustomersecrets.com

For free templates and materials,
visit the exclusive book website:
http://www.threecustomersecrets.com/member

YOU MIGHT FIND THESE USEFUL

CATCH A SLEEP

"The 5-Star Customer Experience - Three Secrets to Phenomenal Customer Service" by Dr Janne Ohtonen, 2017, http://www.threecustomersecrets.com

"Business Process Management Capabilities: A Scientific Edition" by Dr Janne Ohtonen, Amazon Publishing, 2015, https://www.amazon.co.uk/Business-Process-Management-Capabilities-Scientific/dp/9522494372/

"You Think You Are Doing Well? Become a Winner with Customer-Centric Process Leadership! " by Dr Janne Ohtonen, Amazon Publishing, 2013, https://www.amazon.co.uk/You-Think-Doing-Well-Customer-Centric/dp/952680550X

"Outside In – The Power of Putting Customer at the Center of Your Business" by Harley Manning, Amazon Publishing, 2012

"The Outside In Corporation – How to Build a Customer-centric Organization for Breakthrough Results" by Barbara E. Bund, McGraw-Hill, 2006

"Strategy from the Outside In – Profiting from Customer Value" by George S. Day, McGraw Hill, 2010

"5 Star Service – How to Deliver Exceptional Customer Service" by Michael Heppell, Pearson

"What Customers Want" by Anthony W. Ulwick, McGraw-Hill, 2005

"The DNA of Customer Experience" by Colin Shaw, Palgrave Macmillan, 2007

"Extreme Trust: Honesty as a Competitive Advantage", Penguin, 2012

"Customer Expectations: 7 Types all Exceptional Researchers Must Understand" by Scott Smith, Ph.D. December 10, 2012 at http://bit.ly/cust-expect

BACK TO SCHOOL

Customer Experience Management brings real business results through optimising customer interactions to maximise the client lifetime value. A competitive advantage no longer comes from superior products or services only; it comes from personalised experiences: customers want to get what they want, where they want it and when they want with a smile.

Many businesses are struggling with this change ripping through the business world, because of old industrial-age mind-sets. The first step to managing your customer experiences is to understand your client at a deeper level. For this purpose, we created Customer Experience Blueprint.

This course allows you to create a systematic and profound analysis of your customers, their expectations, and psychological needs so you can boost revenue, decrease costs and improve customer service. This client experience management course contains powerful tools, techniques, and teachings to deepen your knowledge of customer emotions and other relevant aspects.

To simplify the course even further, templates and videos are provided with that guide you every step of the way, saving you time and energy when putting together your customer experience strategies. After creating your first Customer Experience Blueprint with the help of the material this course provides, you will have a reliable roadmap to improve your business results continuously. This makes you a highly valuable member of your team and company.

---

*Join the course and start creating your own
Customer Experience strategies TODAY!*

https://www.udemy.com/customer-experience-management-blueprint/

INSTEAD OF NETFLIX

Visa Customer Service Parody
http://bit.ly/ceb-parody

Financial Linkage - Demonstrating CX Value to the CFO
https://www.cxpa.org/viewdocument/cxpa-sponsored-webinar-by-medallia

Changing the Organization to Deliver Solutions - Interview with Dr Gulati
http://bit.ly/ceb-gulati1

Strategy from The Outside-In - Interview of Professor Day
http://bit.ly/ceb-outsidein

Case Story: United Breaks Guitars Song
http://bit.ly/ceb-united

How Understanding the Customer Experience Saves More Than $800,000 by Dr Ohtonen
http://bit.ly/ceb-800k

Radio Interview of Dr Janne Ohtonen by Bill Black (Exit Coach Radio) on Customer Experience Strategies
http://bit.ly/ceb-black

Customer Experience Case Study: Aviva Halley & The Coopers
http://bit.ly/ceb-aviva

Excellent Customer Service
http://bit.ly/ceb-service

Customer Loyalty
http://bit.ly/ceb-loyalty

Why Customer Experience Matters - in Numbers
http://bit.ly/ceb-numbers

Car Breakdown - a short video about great customer experience
http://bit.ly/2dnq3xw

TOILET READING

Openet Learning Centre
https://www.openet.com/learning-centre

Customer Centricity – How to Move from Talk to Action
http://customerthink.com/customer-centricity-how-to-move-from-talk-to-action/

Can Voice of Customer Mislead Our Business?
http://customerthink.com/can-voice-of-customer-mislead-our-business/

Does the Brand Promise Matter for The Customers?
http://customerthink.com/does-the-brand-promise-matter-for-the-customers/

Messy processes – Messy customer experiences?
http://customerthink.com/messy-processes-messy-customer-experiences/

Is this the Moment of Truth Everyone Is Talking About?
http://customerthink.com/is-this-the-moment-of-truth-everyone-is-talking-about/

3 Reasons Why Customer Complaints Are Your Friends
http://customerthink.com/3-reasons-why-customer-complaints-are-your-friends/

5 Things Every Executive Should Do to Build Customer Experiences
http://customerthink.com/5-things-every-executive-should-do-to-build-customerexperiences/

3 Principles That Turn Mediocre Customer Experiences into Remarkable
http://customerthink.com/3-principles-that-turn-mediocre-customer-experiences-intoremarkable/

Link Operational Metrics to Customer Loyalty Metrics for Better Financial Outcomes
http://customerthink.com/link-operational-metrics-to-customer-loyalty-metrics-forbetter-financial-outcomes/

PEOPLE WHO CAN HELP

Since its formation in 1999, Openet has successfully established itself as a trusted advisor to service providers in the telecommunications industry. We have carried out more than 160 transformational projects for 70 clients in 65 countries.

Our consultants have expertise covering the main layers in the telecommunications stack such as Access, Core, Charging, Rating, Billing, Policy into OSS/BSS functions and digital marketplace across multiple mediums (e.g. Fixed, Mobile, Wi-Fi).

Openet Consulting delivers architectural, operational expertise and services that enable monetisation, service innovation, subscriber engagement. There are lots of initiatives getting service provider attention as they look to streamline operations and run a network with ever reducing Capex and Opex. Operators are re-thinking their position and considering what their 'purpose' is and which companies they are really competing with and how.

Visit Openet website at http://www.openet.com

FOR TECH JUNKIES

Openet provides the solutions and consulting services that service providers need to fast-track their digital journeys. Our real-time solutions provide a digital platform for service providers to be more agile and innovative. This improves how they engage with their customers to drive new revenues and to increase their share of their customers' digital spend.

Since its foundation in 1999, Openet has been at the forefront of telecoms software development and innovation. Its success is personified by the many long-term relationships it has fostered with the largest, most progressive, and demanding operators across the globe. For more information visit www.openet.com.

We believe that it is not always the strongest of the species that survives, but the one that is most adaptable to change. This holds true for our customers and our own business.

We believe in our expertise and our people, and that collaboration raises our performance above others. We believe in open networks and systems as these enable our customers to innovate, adapt and transform the customer experiences they provide. We deliver software solutions and services that reflect these beliefs.

Visit Openet website at http://www.openet.com

FINNS WHO LOVE NON-FICTION WRITERS

This book was supported by the Association of Finnish Non-fiction Writers (AFNW), which is Finland's most prominent writers' organisation.

It is a cultural and professional organisation dedicated to protecting the interests of non-fiction writers. It promotes Finnish non-fiction, protects the copyright and financial interests of its writers, provides support for non-fiction writers and furthers the professional competence and ethics of its members.

The Association was founded in 1983, as a joint organisation for writers of non-fiction and textbooks. It now has over 3,100 members – writers of non-fiction for the general public, textbooks, manuals, dictionaries, articles, etc.

Contact Details:

The Association of Finnish Non-fiction Writers
Mariankatu 5, 3rd floor
FI-00170 HELSINKI
FINLAND

Telephone + 358 9 4542 250
Email toimisto@suomentietokirjailijat.fi

Visit the AFNW website at
http://www.suomentietokirjailijat.fi

FOR DO-GOODERS

Christians Against Poverty (CAP) exists because nobody should be held hostage by debt and poverty. However, the truth is that both these things are rife in the UK. In a society where people live behind closed doors, thousands are desperately poor. Unable to feed their children, incapable of paying to heat their homes in winter, the grip of poverty is relentless. It breaks families apart and drives many to think that suicide is the only solution.

We are passionate about releasing people in our nation from a life sentence of debt, poverty, and their causes. Through our services, which are all run through local churches, we are tackling poverty head-on. Our vision is to bring freedom and good news to people in every community through CAP projects.

"I am overwhelmed by what God has done. To see thousands of lives changed every year is truly wonderful. I do believe that God has given us a 21st Century answer to one of the most pressing social needs within society today. Jesus met people's needs with love, compassion and practical help. Our desire is to simply do the same and watch the miracles unfold. Please get involved in this amazing, God-inspired ministry."
— John Kirkby, Founder and International Director

Our services are offered entirely free of charge because the people and churches who fund our work care about people in our communities who are suffering and want to help. We do have a solution to the debt problems that many are facing, and it breaks our heart to know that people are going through misery when help is at hand.

Part of the pre-tax proceeds from this book is donated to CAP charity.

Read more about CAP at https://capuk.org

RANT AND RAVE

I hope you have enjoyed reading **Marry Your Customers! – Customer Experience Management in Telecommunications**. I want to make the future revisions of this book even better and value your input on how to do that.

Please, send feedback and testimonials for this book to janne@threecustomersecrets.com

---

Anything you submit as a comment or testimonial may be used by the author for the promotion of the book or for illustrative purposes. By submitting your comments and testimonial to the above email address, you agree to the Glamonor policy and give the author permission to use and to edit any comments, testimonials or statements submitted. Email addresses collected through the book website may be used by the author or Glamonor to send more information or offers. They will not be given or sold to 3rd parties.

www.ingramcontent.com/pod-product-compliance
Lightning Source LLC
Chambersburg PA
CBHW071208210326
41597CB00016B/1718